A SUCCESSFUL WIFE

Theresia Makau

A SUCCESSFUL WIFE

Keeping Love Burning

Paulines Publications Africa

A SUCCESSFUL WIFE

ISBN 9966-08-245-x
Year of publication 2007
1st reprint 2011

Nihil Obstat: Fr Pelin T. D'Souza
Censor

Imprimatur: H.G. Raphael S. Ndingi Mwana 'a Nzeki
Archbishop of Nairobi
Feast of St Matthias, Apostle
14th May 2007

PAULINES PUBLICATIONS AFRICA
Daughters of St Paul
P.O. Box 49026
00100 Nairobi GPO (Kenya)
E-mail: publications@paulinesafrica.org
Web-site: www.paulinesafrica.org

DISTRIBUTION CENTRES IN NIGERIA

PAULINES BOOK AND MEDIA CENTRE
P.M.B. 21243 IKEJA, LAGOS STATE
Tel. (01) 7741636, 080 66632310
E-mail: paulinelagos@gmail.com

PAULINES BOOK AND MEDIA CENTRE
Kado Kuchi, Along Banex Gwarinpa Express Rd
ABUJA (NIGERIA)
Tel. 080 74795539, 080 74794730
E-mail: publicationsabuja@gmail.com

PAULINES BOOK AND MEDIA CENTRE
P.O. Box 9763, G.P.O. 10 GARKI (ABUJA)
Tel. 080 30535125
E-mail: paulinesabuja@hotmail.com

Cover design by Frederick Katulu
Layout by John Agutu

Paulines Publications Africa is an activity of the Daughters of St Paul, an international religious congregation, using the press, radio, TV and films to spread the gospel message and to promote the dignity of all people.

Dedication

To my husband and the children,
for their special love.

Acknowledgements

Paulines Publications Africa
for the permission to use copyright materials and
all the couples who talked about their marriage.

CONTENTS

PREFACE

This book is for young women just preparing for marriage. It is also for newly married women. And yet, the veteran wife is sure to find one or two ideas to incorporate into her already established order.

The inspiration to write this book came during extensive discussions with people young and old. Many of them felt very strongly that marriage was not relevant in today's world and that it was fashionable to get into free unions. I should like to tell them that properly understood marriage can bring profound joy to the spouses. But when misunderstood and entered into hurriedly as a result of mere infatuation, it can also cause profound misery.

It has also been my observation that there are men and women who do not understand the intrinsic psychological differences that make their spouses behave the way they do. This misunderstanding often causes undue stress to the relationship and may even lead to divorce.

The security of children has also been given due attention, since most parents are rightly apprehensive about the monster lurking in the dark in the form of paedophile who may molest or attack their children. Such monsters are usually not easy to discern, as they appear in society and in the neighbourhoods as perfectly normal people.

I discovered along the way that public relations tactics are also required in marriage for the spouses to live together till death does them part.

It is my hope that all will find something to add to their knowledge about this noble but often misunderstood institution of marriage.

All the experiences narrated in the book are real, but in most cases the names have been changed. A few people have allowed the use of their actual names wherever indicated.

INTRODUCTION

'Success' means different things to different people. Some view success in terms of wealth and power. Others look at it less in terms of how high they climb in the corporate world and how much money they make but more in terms of the quality of time they spend with their loved ones and what type of families they bring up. They realize somewhere along the way that although money is important in keeping the family happy it is not everything. A successful wife puts aside many things and strives to achieve goals and objectives pertaining to marriage.

A successful wife is a combination of factors such as attitudes, discipline, effort and style. But when a man gets married, it is because of certain qualities that he finds in the specific woman that he chooses. He leaves all the other women in the world and marries this particular one; because of who she is and not of what she has. It is the hope of every man that these noble qualities he admires in the woman he marries do not change afterwards. However, it has been said that the realistic husband knows that these qualities, be they physical or behavioural, are bound to change as age catches up. A successful wife holds her husband in dialogue on the sensible expectation on such changes and at the same time endeavours to maintain the early qualities or improve on them as far as is practically possible.

When a woman gets married she does so with certain goals and objectives in mind. The hierarchy among these goals may differ for different women but their nature is about the same for most of them. A woman wants love, companionship, security, understanding/trust and above all a family, and with children.

How Does a Successful Wife Set Out to Achieve These Objectives?

A successful wife does things that set her apart from the rest. It has been said above that every woman desires love, companionship, trust etc. The question then remains; are these sufficient and necessary to declare her successful? And if so, who judges?

The question of who judges a successful wife in her marriage remains controversial. It is controversial because she can decide to judge herself, and evaluate how many of the objectives she set out to achieve have been accomplished. But society also judges her, and the list of people and their expectations can be long.

Irrespective of who judges, it is important to remember that hers is a journey. It is not a state. It is a journey from courtship and marriage through bearing and training children to her old age. Anyone who sets up a business periodically evaluates its performance. Similarly the wife must occasionally ask herself whether she is attaining her objectives. She may occasionally have to analyse her situation and make the adequate adjustments.

Women ought to be first and foremost feminine, and it is femininity that sets her apart. She is kind, caring, nurturing, consoling, friendly, and forgiving. These are the qualities where woman's strength lies. Her success in marriage depends on how she combines and uses these qualities.

Let us keep in mind that a wife operates within the institution of marriage. Every institution has its own set of rules and regulations that govern its efficiency. It also has a hierarchical structure, where the head enjoys a certain amount of power. There can never be two heads in the same organization with equal power. They would tear the institution apart. The successful wife knows that while her husband is the head of marriage, she is the heart of it. If the heart does not function, the body will not work.

It is by understanding the structure and function of the heart that a successful wife sets out to achieve her objectives. This entails combining various factors.

What Are These Factors?

These will include: commitment, communication, forgiveness, orderliness, diligence, adjustment, love, understanding, conflict, resolution skills…

JASSY AND DENNIS

Jassy, 28, was teaching chemistry at the Provincial Girls School of the Rift Valley Province of Kenya. Her mother was the principal. The family lived in the school compound. Her father was the principal of a neighbouring boys' school, commuting daily to and from it.

Once during the school holidays many people gathered at the Girls' School compound for a wedding reception of one of the lady teachers. The music was sweet and many people were dancing. Jassy was seated with her parents and discussing the event, when Ray came over and asked her to dance. She quickly joined him to the dance floor and discussed politics and education. They found dancing to such an old tune rather funny and they laughed. Ray worked in the same school as Jassy. They had known each other for the past five years but just as good friends.

Jassy's mother, Jane, whispered to her husband, Joe, that Ray looked a fine lad and could make a good husband. 'When will you ever stop match making?' Joe reproached her and changed the subject.

The bride and groom had danced to a few tunes and left for their honeymoon without announcing their departure. Many people thought they were still around. However, so many were preoccupied with their own problems and worries that they did not notice their absence.

Soon the music was over. Ray escorted Jassy back to her seat and excused himself: he had work to do. Jane whispered to her daughter, "He looks a very fine man, he can make a good husband'. 'But we are only friends, Mum', Jassy protested. She did not like her mother's insinuations about her unwed state. Presently, Jassy's parents also excused themselves and left. Jassy looked across the dance floor and saw her friend, Koki, in deep conversation with a man who seemed love smitten. She wondered why people fell in love and how it felt. She had never been in love and always thought it was not for her. She liked her life just the way it was.

Dennis was a government undercover operative based at the Rift Valley Provincial Headquarters to track the movements of a gang that was behind the tribal clashes in that area at the time. It was an extremely dangerous operation because members of this gang killed people mercilessly and robbed homes and shops. He prayed that the operation would end soon so that he would go back to his normal work.

Denis was tall, strong and handsome. He had studied Architecture and Law at the university and worked for a private firm. Tonight he had not been invited to this party, but he was tracking somebody that he knew had needed information.

He stayed in the background because he did not want to make his presence obvious. However, from behind the scene he could see all that was happening and who talked to whom. He looked across to where Jassy was seated and noticed her sage-green long dress that covered her feet to the ankle. He noticed her pretty face and the beautiful hair combed back away from her face. Something about her disturbed him. He was sure he had not seen her before yet he did not want to go away without talking to her.

Jassy was thinking about leaving when suddenly in front of her stood this tall, strong and handsome man. "Lady?"

Denis said as he offered his hand. Jassy looked up at him a bit surprised at this man who suddenly appeared from nowhere. She was sure she had not seen him before but he was the best looking man she had ever seen. There was something about him that she had never seen in anyone else.

"Hi", she finally managed to say after studying his face for a while. "Shall we dance?" Dennis asked and prayed to God silently in his heart that she would not say no. "I would love to dance with you" Jassy said without hesitating.

Dennis knew how dangerous it was to expose himself so much and the trouble he could be in with his boss for it. He promised himself that he would only dance to that one tune and disappear. He had been so long in the bush that he missed the civilized world.

As they talked, that tune ended and they had to sit down. They exchanged contacts but before they parted another slow tune started and Dennis led Jassy to the dance floor. Being in each other's arms felt so good that they really did not dance. They swayed in each other's arms and each prayed silently that that moment would last forever. Dennis looked down at her face and she looked up at his. Their eyes met and locked. It was a magical moment. Everything else around them seemed to melt away until it was only the two of them left. The passion that burned between them was something out of this world. Neither of them tried to fight it. But the music came to an end and the DJ announced a commercial break.

Two years later Jassy and Dennis were on the same grounds but now as bride and groom. Jassy loved Dennis so much that she was ready to live with him even if he lived in a ditch full of scorpions as it were. Dennis felt the same and after their honeymoon, she only wanted to live with him alone and to spoil him. They actually spoilt each other as much as they could.

Jassy and Dennis had gone through a turbulent courtship. The parents of Dennis were completely opposed to the idea of Dennis, their first son, getting married to a girl from a different tribe. Jassy's tribe had been locked in tribal clashes with Dennis' for a long time. Dennis' mother was also not happy with a girl who had had so much schooling and was working. She would have preferred someone who would stay at the village with her and help in fetching water, firewood and milking the cows in the morning. Of course this was not acceptable to Dennis and he was not going to be separated from his love by anyone or anything. He enlisted the help of Jassy and they were able to convince his parents that his getting married to Jassy did not mean abandoning them. He would still visit them and their neighbouring tribe would no longer be their enemy but their friends since they were going to be relatives.

Jassy and Dennis had the happiest honeymoon one could ever imagine. They had gone through pre-marriage counselling and knew what to expect and how to respect each other to keep their love burning.

Jassy transferred from the Girls' School and moved to teach in a City school to be next to her husband. At the new School she was lucky to be housed and her husband moved there after completing his assignment in the bush.

Since Jassy had studied up to Master's level of education, she thought she was lucky to be at the big city and could therefore enrol for evening classes at the local university for a Doctoral programme. Dennis did not take such a proposal very kindly. He said he wanted to come home in the evening and find her there but not to wait for her to come and cook for him 'after midnight.' He actually made it very clear that she should have done all the studying before getting married. Jassy thought Dennis was right about studying and starting a family at the same time and shelved her study plans for the future.

Jassy and Dennis found pleasure in going to the lakeside, rowing boats together, mountain climbing and generally being outdoor. Most Saturday afternoons they went to Uhuru Park and chatted away the afternoons watching people rowing boats and occasionally rowing also and enjoying the cool breeze.

Jassy and Dennis were so busy and content with each other that soon two years were over without a child. Their parents started showing concern, and a 'friend' even advised Dennis to quietly abandon Jassy and get a woman who would give him children. Dennis did not take such advice but he organised for a medical check-up for both of them. The doctor did not find anything wrong with either of them. And as if the baby were waiting for the presence of the doctor to get formed, Jassy's pregnancy test was positive the following month.

The babies came quickly, one after the other. In five years they had three children. With the arrival of the babies also came the arrival of the green-eyed monster – jealousy. Dennis started feeling that the attention he used to get from Jassy was waning. He complained that Jassy was spoiling the babies by being with them all the time when she was at home.

Jassy noticed this and quickly remembered what the counsellor at the pre-marriage classes had told them. "Let the children fit in your and your husband's plans and not the other way round." She also reminded Dennis that her love for him came before that of the children.

Jassy also suffered episodes of jealousy. At one time Dennis was bed ridden and had to be cared for like a baby. Jassy got a girl to look after him while she was at work. One day she came home at lunchtime and found him laughing heartily from the story this girl was reading to him. Jassy panicked. "How could this girl make Dennis so happy? What else is she capable of?" Jassy made a mental note to change this girl as soon as possible.

Jassy at this time was also fighting her battle with her own mother and mother-in-law. Both women had retired from active life and wanted to spend as much time as possible in Jassy's house. Whenever they were there they felt that they knew better than Jassy what was best for the children and therefore used to give conflicting instructions to the maids from what Jassy had given them. They also tried to put Jassy down and told her that she was not doing enough for her husband. Her mother-in-law would complain how her son liked food cooked in a particular way and that Jassy could not do it.

Jassy was near the end of her tether. She organised for the two mothers to join women groups where they could get occupied and give her a breathing space. She also enrolled the maid in a college that offered boarding facilities to study house keeping and laundry and got a more mature one. Dennis wondered what happened to the maid who was so responsible and made him so happy when Jassy was away. Jassy reminded him that the girl had come to work for them only temporarily while waiting to go to college and that she was already in class.

The more the children grew up, the more Dennis grew to like them. He eventually got so attached to them that Jassy started complaining that he was spoiling them. They did not need counselling for this. They knew what to do.

Jassy allowed Dennis his freedom to be who he was and he gave her freedom too. They trusted that what they had was too precious to be spoiled by suspicion. Both did their part and left the rest to God to take care of. Their faith helped them to cope with the most trying moments of their lives.

Today Jassy and Dennis are grand parents, and they are careful not to overly interfere with their children's lives. They are making plans for survival when one spouse dies.

COLLEEN AND ROGER

Colleen met Roger at a party organized by common friends. They were both visiting a neighbouring country for business. Colleen was a lawyer and Roger an engineer. They exchanged contacts and when they met a second time in a different country they were sure that they were meant for each other.

Later they got together and within six months they were married. They were friends and not lovers. They discussed marriage without love and thought it was all right. After all, who said that love existed? Roger kept repeating this all the time when the issue came up for discussion.

Within two years they had two children and this forced Colleen to scale back on her career and look after the family.

Colleen grew to like Roger very much but she did not study him when they were engaged. She did not notice his quick temper and a mouth as sharp as a razor blade when provoked.

When Colleen stayed at home she unfortunately did not look after herself properly. She allowed herself to get overweight and slovenly. Roger became withdrawn and hardly made love to her. Whenever they argued he would call her a pig and once he slapped her. Once when making love to her he mentioned the name of another woman.

Roger confided to one of his friends that he could not stand Colleen any more. He hated her and her children. One day he packed his things and left for good. He was contemptuous.

If Roger had not become contemptuous, he would have sought therapy and they both would have been helped.

Their marriage would most likely have been strengthened. He blocked dialogue by being contemptuous.

Reflection

Although Jassy and Dennis faced problems like jealousy, in-laws' interference and a major sickness in their marriage, their marriage did not fall apart. Instead, their love was strengthened. By the time they reached old age, they decided to stick together on their own rather than moving to live with their children and probably interfere with the way they did things.

Colleen and Roger on the other hand do not seem to have met the type of problems that Jassy and Dennis met. Yet their marriage fell apart.

How did Dennis and Jassy deal with jealousy? First, they realised that a bit of it was there, but it was not a major problem. It was a minor problem and they tackled it as a team. Sometimes a bit of jealousy is important in a relationship for each to realize the treasure that the other is. However if not guarded it can ruin the relationship. It should therefore not be allowed to grow.

Dennis and Jassy treasured their relationship and they treasured each other. The love they had for each other was so strong that Jassy was ready to live with Dennis in a ditch full of scorpions. They were ready to sacrifice for each other.

Their relationship was unique, Jassy was special to Dennis and Dennis was special to Jassy. Therefore they kept their fidelity.

Jassy and Dennis also knew that their relationship was indissoluble and that they had to work to keep it that way. When they were pronounced husband and wife they swore in the presence of God and the priest to live together "till death do us part". They were not going to let somebody or something separate them.

They were also happy about the fruitfulness of their

union. A union between a man and a woman should be fruitful. This is one of the main purposes of marriage. The fruits of a marriage are the children that God blesses the couple with. Sometimes a couple may opt not to have children because children are a heavy burden. This is really an unfortunate attitude. God's idea of marriage was that it would be fruitful and permanent. One should however not be insensitive to those who would like to have children but cannot.

Colleen and Rogers's marriage lacked commitment, both to each other and to the vows they made before God and witnesses. They forgot that if marriage was good enough to enter into, its goodness should have compensated its responsibilities. Their union was superficial and lacked depth. They did not go for pre-marriage counselling. They probably thought that was for people less intelligent than themselves and that they were in control. They seem to have excluded God from their relationship completely and did things with their own strength and power.

CHAPTER 1

PREPARING FOR MARRIAGE

What Makes a Successful Wife?

'She has problems, but does not procrastinate or avoid them; she solves them judiciously.'

She puts in place attitudes, disciplines and effort that spell success in whatever she is going to do.

Just as a sportsman who wants to win a race makes adequate preparations, and as the architect who wants to put up a permanent house lays strong foundations, so it should be for a woman who wants to succeed in marriage. No one in one's right mind would wake up one day and decide to drive a car or fly an aeroplane without having been inside one before. It would be suicidal. Yet many young people are doing exactly that with their lives by rushing into marriage with complete strangers. How can they expect it to work?

One does not get into marriage simply because "society has been watching and now she is approaching 30 and no suitor has come by, therefore she must get married to the next man she meets." Society can put pressure on the young people in very subtle ways, which suggest that years are getting by and that one should do something about it. The girl who makes a successful wife does not bend to such pressure. She remains confident that one day the right man will come her way. Finding the right husband involves falling in love and ascertaining that this love is true and not an infatuation. While in love, there are certain issues that must be considered and understood before marriage. These include: medical history of the spouse, the past life, finances, cultural background etc. These must be considered because of their heavy impact on marriage.

Past Life

This is the history of the person. What has he been through? Was he married before? What factors affected him when growing up? Is he an orphan? Is he from a single parent, or from a broken family? Is he from a dysfunctional family? Is there a history of violence at home? How were his parents: authoritative, authoritarian or permissive? Does he love? What values does he hold?

The home where the individual grows up plays a major part in shaping his personality, therefore it is important to get as much detail as possible about it. These details can be obtained either by asking him, or his relatives and friends, by casually discussing the past, but without getting too nosey.

Medical History

Some medical conditions pose a real challenge in any relationship. Any form of illness is a challenge even to the most committed families. Therefore there are conditions one must understand clearly, so that when deciding to marry the condition will not come up as a shock and destroy the relationship.

The list of these conditions can be long, but the most important ones include:

Acquired Immuno Deficiency Syndrome (AIDS)

The Guide to Medical Cures and Treatments says that AIDS has become a major problem in the world in the last two decades or so. Anyone preparing to marry must understand this plague and its implications.

This virus is ever changing and attacks the cells that should help the body to fight infection. This virus can get into the body of a healthy person through sexual intercourse, by injecting drugs, using a needle that has been used by someone who carries the disease, through blood transfusion with blood that has not been properly screened or transfer of the virus to the child before or during birth by an infected mother.

The virus destroys the cells that fight diseases in the body system and leaves one exposed to infections of all types, including cancers, skin diseases, lung diseases, fatigue, recurring diarrhoea, weight loss, and even tuberculosis.

Although an infected person may not show any of the above symptoms, he is still a carrier and can pass on the disease to unsuspecting people.

So far there is no known cure for the disease and medical manuals give lists of precautions against infection. This list includes abstinence from sex and avoidance of needles that have been used by contaminated people.

If one is informed of this condition and still decides to go ahead and marry the person who is suffering from HIV, then thy need to visit a heath provider who will furnish them with details on how to live with the disease.

Mental Health Disorders

The Merck Manual of Medical Information lists a number of mental illnesses. It is important for those preparing for marriage to be aware of these so as to make informed choices.

The list includes posttraumatic stress, obsessive-compulsive disorder, depression and mania, eating disorders, sexuality and psychosexual disorders, disorders of sexual function, personality disorders, Dissociative disorders, schizophrenia and delusional disorder, drug dependence and addiction.

It is not possible to give all the symptoms of these conditions but it is important to point out how one can identify an abnormal behaviour in a spouse. Ignorance of some of these conditions can give great pain to the uninformed spouse and can be a cause of an unhappy or failed marriage.

The family physician should be in a position to advise on the threat posed by these conditions but it is better to have some idea of how to identify the most obvious.

Often when the young people are in love they may get blind to these symptoms, which sooner or later will surface. Let us look at some.

Drug Dependence and Addiction[1]

Addictive substances and drugs include alcohol, cocaine, heroine, narcotic painkillers and sundry prescription drugs. Sometimes it may be difficult to tell someone who has a problem with drugs but there are clues or specific patterns of behaviour that point to a leaning towards drug or substance addiction. Some are listed below.

1. The substance is taken in larger amounts than intended by the person.
2. The substance is taken for a longer period than intended by the person.
3. Dependence on sleeping aids and anti-anxiety drugs.
4. The person may become extremely depressed and anxious.
5. The person may spend a lot of time trying to get the substance, taking it or recovering from its effects.

6. The person continues to use the substance even when he knows it is causing problems.
7. There is marked tolerance, meaning that increasing amounts are needed to achieve the same results. For example, someone who got intoxicated after taking one bottle of liquor now requires five to get intoxicated.

Schizophrenia and Delusion Disorder

Schizophrenia is characterised by loss of contact with reality (psychosis), hallucinations, delusions, false beliefs, abnormal thinking and disrupted work and social functioning. It is a major public health problem throughout the world.[2] It is said to affect one person in every 100. Some schizophrenic sufferers may be so impaired that they may not hold a job and require constant care. Other forms are mild and easier to cope with.

Consider the case of John, (not his real name) a former professor of chemistry at a local university. John had a normal youth and adolescence. He was intelligent and was doing well for himself when disaster struck. He started by accusing his wife of having affairs with other men and he even told this to friends and relatives. They felt sorry for him and even some offered to mediate between him and his wife Jane. Jane was shocked at the accusations and devastated by them.

Things got worse when John started accusing Jane this time of trying to kill him. He complained of seeing heavy lorries following his car as he drove home from work. This made him drive so fast as to cause accidents.

Although John received psychiatric help, his condition worsened and he could not continue with his career. Today he spends most of his time in his bedroom where he says he communicates with 'good spirits'. He believes that these spirits bring him messages of 'goodwill' and that they should not be interrupted. Sometimes when people knock at his door, he gets very agitated and complains that they are sending his 'good spirits' away. John does not attend social gatherings because people in those gatherings might contaminate him with evil spirits.

John's case is a severe form of schizophrenia. There is also a mild form, where the sufferer experiences illusions of grandeur or persecution. One may imagine oneself a great politician or a leading personality in his field of specialization.

The case of Kako illustrates the point.

Kako is a beautiful woman who holds a master's degree in medical science. She got married to Kalao and they had two children. Kako started suspecting her husband of having extra marital affairs when she was at work. Her husband also worked in an office but she imagined he was not fully occupied therefore had time to run around. This form of suspicion led to divorce and when her husband left with the children, Kako started accusing her own mother of having been responsible for the breakdown of her marriage. She accused the mother of sending jinx to disrupt her marriage. Kako got married a second time but this time made sure her mother would not go anywhere near her new husband and children. At one time Kako's mother made the mistake of visiting her daughter in the house. When Kako got home and found her mother in that house, she combed it through trying to look for evil charms concealed under chairs, in flowerpots or even inside the flowers in the garden. She told her mother to her face that she did not wish to see her ever again in her house.

Cultural Background

For married couples of diverse cultural backgrounds the danger is real to blame their cultures of origin when marital problems set in. The husband of a woman who picks her nose with her fingers or chews with her mouth open may blame her cultural background while their problems may actually originate in unresolved emotional conflicts carried over from his youth. The cultural backgrounds should be clearly understood so that culture is not blamed for unattended emotional conflicts. All of us are a product of the traditions and customs of the society that brought us up. Whether we like it or not these traditions affect us somewhat.

Knowing the past also means knowing whether he was involved in drugs, violence, incest, or other unions. These vices are not compatible with a happy marriage.

Bride Price

Intricately tied to culture is the question of bride price. In the past when this practice was started in Africa, it was meant to be an exchange of gifts between the two families. It was not meant to 'sell' the girl. However, because of greed, many parents of the girl today ask for huge sums of money before they agree to give their girl in marriage. A girl who is ready to succeed in her marriage should not let her parents 'sell' her as it were. Marriage is supposed to be a free union between the spouses. In Indian traditional marriages, it is the girl's family that pays the dowry. The parents of the girl get together with those of the boy and they decide that a reasonable amount should be paid to the boy's parents before the wedding.

The question that lingers in everyone's mind is: Will this bride price assist in stabilizing the new family unit? If not, why pay it? Most parents nowadays agree for the couple to marry on most issues. If the girl likes to be 'bought', then the parents have no choice.

Finances

If a couple is to avoid unnecessary stress, they cannot afford to overlook this area. They must be assured of a steady source of income, and who of the two will do what to generate this income, or whose income will meet what expenditure. Those who get into marriage without planning their finances, find themselves unable to avoid interference from their parents, who will have to continue to support them.

There are parents unable to let go of their son/daughter, and who use financial inducement as a carrot to control the new couple. The new couple should reject such offers and struggle on their own. Even if they are not employed there are many things they can do from home to generate income. If for any reason they agree to accept financial assistance from their parents, it should be treated as any loan, with plans to pay it back as soon as possible.

Common Interests and Values

People marry to live together and participate together in many activities. If he likes rock and mountain climbing during his free time while she likes reading and watching movies, they might face trouble planning their holidays. These interests must be discussed and understood.

Their values must also be compatible. If he likes many children and she likes only a few there will be trouble. They must decide in advance whether they are willing to commit themselves to marriage without looking back and give it their all.

What human values does he cherish? Are they compatible with hers?

Religion

The ideal is for the two to practice the same faith. Many problems arise when people overlook the issue of faith before marriage. What religion will the children follow? "Difficulties in mixed marriages arise from the fact that the separation of Christians has not yet been overcome. The spouses risk experiencing the tragedy of Christian disunity even in the heart of the home" (CCC 1634).

CHAPTER 2

HINTS FOR SUCCESS

a) POSITIVE FACTORS AND QUALITIES/ ATTITUDES OF A SUCCESSFUL WIFE

Commitment and Communication

On undertaking any venture, the commitment and endurance put into it determines its success. A successful wife gets into marriage without any thought of ever going back. For her it is make or break, completely confident that she will succeed. With this confidence she will do everything within her ability to make marriage successful. Many communities here have a special gift for a girl who is getting married: a bed. The bed symbolises her detachment from her parents' family. She must go out and make her own home. This tells her also that she should make every effort to sort out whatever problems she might encounter in her marriage and solve them. Apart from the symbolic meaning, it means that girls from this community are trained to deal with the sticky issues in marriage and therefore avoid divorce. It serves as a call to devotion and loyalty to marriage. Commitment may mean different things to different people. It means one thing to a wife who is working towards success: *total self giving.*

A local politician once remarked, "Where there are heated arguments, there is peace." Where much remains unsaid, or hidden, there can be no peace.

Peace in a home comes about when spouses openly discuss any issue that threatens to cause discord. Psychologists say that issues that remain hidden in people's hearts or minds cast a long shadow.

A successful wife knows that in a family there has to be open dialogue for the spouses to understand each other. Where there is no discussion, there is trouble. Where the discussion is an outpouring of emotions fanned by unresolved conflicts, there is also bound to be trouble.

Communication in marriage has been defined as 'uninterrupted flow of information between the spouses'. It is the flow of loving thoughts and feelings from one to the other. In an uninformed relationship, this flow can be interrupted by anger, hatred, or un-expressed suspicion. Otherwise if informed and not interrupted, each of the spouses grows stronger because of the joy of being loved, secure and belonging. In a secure relationship, talks, hugs, touches and even the holding of hands flow freely.

One can always tell couples that love each other by observing the silent understanding between them. In some Kenya communities they may not hold hands in public since this is culturally considered vulgar but their facial expressions and general poise may reveal a lot.

Suspicion can damage a relationship to the core. Women are said to hold unexpressed suspicion more than men do for fear of attack by the men. A woman might suspect that her husband is interested in another woman, but instead of confronting him she keeps quiet and withdrawn. When the husband asks, "Is anything wrong dear?" she may answer, "Yes, I have a headache," or "I think it's about that time again," meaning pre-menstrual tension. Suppression of doubt like the foregoing can be very damaging to the relationship.

In communication, choice of words is also important. Although we can communicate through gestures and facial expressions, there is a limit to the extent to which these can convey the message to be got across. Our words can hurt, soothe, encourage, discourage or perhaps clarify an issue.

A touch can communicate a message like "I love you" or "I care" but it may not tell him "you are more lovable when clean and in clean clothes." Failure to touch him may not communicate the message "your breath offends" or "your accumulated sweat offends."

Experts in communication recommend the use of concrete words instead of abstract ones. Mention bricks and tables, plates and spoons or even parents and children.

Matters of personal hygiene may be difficult to handle with strangers but with loved ones it is easy to get across a message like "Do your teeth need to be checked again dear?" "Your breath has some odour." Or "another throat infection?, the breath has that smell again."

Words can elicit negative emotions and feelings as much as positive ones.

Consider the following "You are slovenly and unattractive, I cannot go with you to the party," or "I like going out with you but can you please put on something more presentable?"

In most communities it is important to use words such as please, thank you, sorry. In marriage these words are also important. When we offend others and are truly sorry, they understand and forgive us. But when we go about with the attitude "It serves you right" or instead of saying "thank you" we just keep quiet as if to say, "You have done your duty, it was my right."

When we tell someone "Please don't do that again," it hurts. Unless one is a sadist who likes to inflict pain, one stops.

By choosing the correct words to describe exactly what she wants to say, a successful wife communicates the right message and attitude to her listeners. She communicates to her husband that she trusts him completely and it is because of that trust that she chose him out of all other men in the world.

Trust

Trust is at the basis of a lasting relationship. When we trust we believe that we can rely on the strength and goodness of the trusted one. A wife who has seen her sisters abused and trashed in marriage may question her sanity when she wants to trust her husband. She may ask herself: what guarantee can there be that it will work with me when it has failed with so many others? These are important questions, since we live in a society where people have witnessed misery in the institution of marriage. But let's be realistic, not all married women get abused and the abused ones are not even the majority. It has been said that 'Parasites may attack a strong and healthy plant, but because of the nourishment from the fertile soil, it grows even stronger. It is so with a marriage built on a strong foundation'.

A successful wife identifies the strong points in her husband and sets out to complement these with hers; thus instead of two weak individuals they form one strong team. These strong points will be what she values most in people. He may be industrious, he may be persevering, respectful of the rights of others, etc. The list may be long. Despite these strong points, he may have the temper of a volcano, or snores like a tractor. A successful wife takes stock of these points about her man and learns how to live with them. She trusts in his essentially good nature. She knows that even when he fails, it is not because he wanted to hurt her but it was just one of those unfortunate life happenings. She has a permanently positive appreciation of him.

Trust has to do with a constant reminder to one that to err is human and that the just man falls many times a day. Did they say seven times?

A successful wife also knows that her husband may repeat the same mistakes because of circumstances beyond his control. She therefore works with him to try and solve the problem together. A good example here is a husband who snores so loudly that the noise becomes a terror. He might be suffering from a condition that requires medical attention. So instead of deserting marriage because of the noise, she helps him to seek medical help after trying different sleeping positions.

Sometimes the idea of divorce pops up on encountering problems in marriage. But running away from problems does not solve them. Many arguments in favour of divorce have been put forward. It has been argued that it is better to divorce than to be stuck in an unhappy marriage. Others look at divorce as an inevitable good point. But what about working at the unhappiness? Since she decided to get into marriage and stay there forever, a successful wife does not allow anything to break her commitment. She looks for solutions in all problems.

She knows that the fact that she became the wife of this man cannot be changed, either in heaven or on earth. The fact that this most flawed man is the father of her children cannot be changed, or erased from reality. It would be like trying to erase history from history books. A successful wife knows that divorce has a cost. The children of divorced parents suffer tragic consequences that surface later in their lives.

"Divorce ends a marriage but does not end a relationship. The cloud of sadness once a love is lost, hovers over the survivors for a long time if not for the rest of their lives."[3] Indissoluble marriage is guided by this principle. A successful wife seeks therapy from an expert on marriage to settle her problems.

Orderliness/ Diligence

"An orderly person follows a logical procedure, which is essential for achieving any goal – in organizing things, using time, carrying out activities on his/her own initiative without having to be constantly reminded."[4] A successful wife has a timetable. She gets up at the same time everyday and carries out her activities of the day according to that plan. She has set time for meals at home, especially breakfast and dinner when it is possible for all or most of the family members to be at home. It is usually during mealtime that the family can be together and discuss important issues about their welfare.

A woman who is orderly finds that she has a lot of time in her hands, because everything is in its place. Consider the woman who has only twenty minutes to prepare breakfast for her family and still go to work. Things will be easier if her sugar, coffee, cereals and jam are kept in the same place and labelled clearly. Picture the same woman who has to look for her coffee from all over the kitchen and eventually find it in the sugar tin labelled 'salt'!

This orderliness will also be transmitted to the children since home is the first school.

"Children have never been very good at listening to their elders, but they have never failed to imitate them."[5] Order is also reflected in the way she conducts herself and her mode of dress. Somebody once said that a well-dressed woman is like a pearl in its shell.

Anyone who sets out to achieve an objective knows that hard work and great effort are a must. Running a family is no mean task. It even gets more complicated when the wife has to run the home as well as work in an office. A successful wife is often the first to get up and the last to go to bed. She occasionally prepares treats for the family. She presents these treats with such love that the family members feel that they are the most special people on earth.

Adjustments

Adjustments in marriage mean several things to the newly married couple: to the wife it means detaching herself emotionally from her parents and relatives and bonding with her husband. To the man it means the same. For each of the two there must be abandonment of whatever ties they had with friends and relatives and a fresh beginning with each other. This does not mean not having friends. Friends are always necessary, but by no means must they be allowed to run the show. A woman who always runs to her mother to ask for advice when the baby coughs is clearly not adjusted. She should seek the help of a medical doctor when the child is unwell and not rush to her mother who is unlikely to be a medical doctor. Even if she were, medical ethics would not allow her to treat her own family.

At the wedding and during the honeymoon most couples are happy, and they hope that this happiness will last forever. But sooner or later they wake up to realising that the romance that brought them together has waned and their former independence is replaced by duties and responsibilities in the new home.

It is during this period of disillusionment that a successful wife braces herself for the challenges ahead. The inability by either spouse to detach emotionally from the parents and to bond with the marriage partner leads to many problems, which may well end up in divorce.

One must understand what marriage is, why she is entering into it and the enormous responsibilities it places on her.

There must therefore be adequate preparations before marriage.

Before marriage the girl who will be a successful wife decides what type of man she would like in her life. She meets many suitors but settles on the only one she loves and whom she feels most compatible with her expectations.

Temperance – Self-Control

Emotions

Emotions are passions. When kept under control they have their value and significance in our bodies. They often cause us to act the way we do. Some emotions are fear, contempt, anger, disgust, surprise, shame, interest, distress and enjoyment.

Experts in behaviour say that emotions can play havoc with our bodies if not kept in check by a deliberate and calculated effort to educate them. Learning to identify our emotions and to control what they urge us to do is the first step towards an acceptable behaviour. The idea is not to suppress the emotions, but to express them in an appropriate manner and not to let them run out of control.

Both lay and spiritual literature is full of examples of people who fail to control their emotions and suffer evil as a result. Emotions or feelings should never be the basis of our decisions.

Temperance is the virtue that helps to control bodily pleasures, for over indulgence in such pleasures can be calamitous. It is a decision to keep our passions in check. Throughout history we read of great men and women who would have done very well or were doing well but over indulgence in these passions destroyed them.

Temperance helps a successful wife to keep her emotions under control. She does not allow lust for sex or gluttony for food to run out of control. When she loves, her love for her family is not just emotional. Her love is a mental attitude and decision to do what is best for herself and her family. She also endeavours to inculcate this virtue in her children.

THE DRUNKEN MOTHER
(A true story)

Kathie loved going out with her husband Kim. Kim's friends were her friends and her friends were Kim's friends. She also went out with her own friends.

Kim would have liked Kathie to stay at home and look after their son but Kathie's lifestyle prevented her from staying in the house. She could not spend a whole day in the house without soon getting bored and calling one of her friends to meet her at a pub.

Kim loved Kathie very much, but was too busy with his work to notice her drinking pattern. She had several workers in the house but did not bother to instruct them on what to do. The workers made major decisions in the kitchen such as what was to be cooked for lunch and dinner and even what special treat would there be for Kim and the young boy for their birthdays.

Kathie came from a well-to-do family but her parents had not really been there for her. Although she had been through school and had a degree in social psychology she did not apply what she studied to her life. She loved leisure and pleasure and lived almost only for that.

It was on a Friday evening, what people call members' day. On such days some people go to clubs and pubs for drinks with their friends.

Kim was on a business trip and Kathie decided to go out with her friends to a popular joint some 50 kilometres from home. Around 10 p.m. Kathie was called from the house by the maid to go and take her son to hospital because he was down with diarrhoea and vomiting. Kathie was already drunk and mumbled something to the maid to the effect that she was on her way back and not to worry about the child. It was not until 1 a.m. that Kathie made it to the house. The maid had not taken the child to the hospital and he was badly dehydrated.

With the help of a neighbour the child was taken to hospital at this hour but it was too late. He was rushed to the intensive care unit but died soon afterwards of measles and a bacterial infection.

A Successful Wife Is Informed and Understanding

Information

Being informed leads to understanding. An informed wife knows the factors that lead to a particular behaviour and how to help people repeat desirable behaviour and discourage unwanted behaviour.

A successful wife is informed. She keeps in touch not only with new developments in her field but also with what affects her everyday life.

She is also informed about relationships and why occasionally even happy couples may fight after a loving episode. At a recent seminar on the family, men gave the following list of expectations from the women they married. They all expected the women to be:

- Beautiful in both body and mind.
- A loving companion despite financial difficulties.
- Ready to persevere in times of difficulties.
- Not fighting to be the head of the family.
- Positive about sex.
- Submissive.

Understanding

Understanding is a gift of the Holy Spirit, as we learned when growing up. A wife who understands her husband knows why he behaves the way he does, and works with him in that understanding to create harmony in the family. When he takes 'space' away from her, she grasps the fact that he needs some time to sort out something that might be bothering him. She also empathises with him when he has a problem that needs sorting out.

Forgiveness

It is inevitable that when people live together they are bound to hurt one another and this hurt may develop into a nasty relationship. A successful wife knows the power of forgiveness when she is wronged. She knows the joy one experiences on being pardoned for the hurt caused to others. She also knows that when we do not forgive, the grudges that we keep can really hurt us. Failure to forgive has been compared with taking poison and expecting the person we are not forgiving to die instead.

A woman who forgives and continues serving her family cheerfully is usually so busy serving that she has no time for self-pity, and rarely falls ill. It has been medically proven that thoughts can make us ill. Therefore the positive thoughts of forgiving and letting go of bad emotions keep a successful wife happy.

When growing up she was always reminded to keep a positive mental attitude. Such attitude is most essential for a successful woman. Guided by this attitude she is able to sort out thoughts and feelings, and also discern and forgive attitudes and motives in those who wrong her.

Forgiveness, however, is not to be confused with passive acceptance of wrongs and doing nothing about it. Whatever is evil has to be condemned in the strongest terms, and measures taken to correct it.

Perseverance

Her mind is set: she is going to be in this marriage till death does them part. She then takes steps to achieve her objectives despite difficulties or discouragement from her peers or other obstacles.

This mental disposition is an aspect of the virtue of fortitude, consisting in enduring difficulties without giving up. Perseverance in marriage means understanding this most flawed husband, and developing strategies for avoiding his weak points and supporting his strong ones. Understanding the flaws that make spouses hurt each other is a boon in marriage.

The Tragic Flaw

Each human being is born with weaknesses, which make people hurt others. Often this hurt is caused unconsciously. When people fall in love they tend to forget about such weaknesses, only to rediscover them in marriage soon after the honeymoon.

Marriage counsellors call this time the disillusionment or discovery period. During it the initial romance wanes and reality waxes. The woman discovers that after all this other human being is real, and has all the weaknesses of the people around her before she fell in love. He snores at night, he has bad table manners, but more seriously he has a strong drive for power and wealth. He on the other hand discovers that she gets moody and withdrawn, she is selfish and conceited or that she is a spendthrift.

This inherent weakness is what Plato, the great philosopher, called the tragic flaw. Others have called it the fatal flaw while yet others have called it the dominant defect. Causes of it may be pride, laziness, drunkenness, lust, insensitivity, tactlessness, selfishness or perhaps envy.

The newly married woman may mistakenly think that this flaw is peculiar to her husband if she has not received adequate pre-marriage preparations and strategies to handle disillusionment.

Plato thought the tragic flaw needed some form of cleansing of the body to purify the sufferer. Moral philosophy advises controlling the lusts of the eyes, the lust of the flesh and the pride of life.

A successful wife tries to control this defect in herself, and helps her family to fight it through self-control. She has seen and read about people who would have been successful but were brought low by this defect.

Exercises in self control mean simple but important things like saying 'no' to a second helping of a tasty meal, to a second bottle of beer, taking a cold shower, suffering people one may find tactless and boring, saying "no" to sex at the wrong time etc.

Jani was happily married to Jila and they were an outstanding couple. He cherished her and she loved him. But occasionally Jani would experience such strong emotions of hate that she wondered how she was going to survive in the marriage. She hated Jila for being still emotionally tied to his mother and for not adjusting and getting emotionally bonded to her.

The relationship between Jani and her mother-in-law was tense and she kept asking herself whether she would survive. Once Jila's mother bested Jani in an argument but he did not go to her defence. Her negative passion for both of them grew into a rage. She wondered how she had been trapped into this. Suddenly she remembered that she was in that marriage for better and for worse and she was not going to let her mother-in-law part them. She resolved to confront her mother-in-law and tell her that her style of doing things was different and she was not going to accept such humiliation. She also realized that pride was her defect and resolved to work at it.

Optimism

Optimism in marriage helps a successful wife to look at the bright side of things. An optimistic disposition is strongly related to the virtue of hope. To hope means not to give in to discouragement and despair. In marriage one may get discouraged by the thought of waking up one day only to realize that the two parents are suddenly alone or only one is left and the children have left to form their own homes. One may feel abandoned and look at life as a sham.

The virtues of hope and optimism keep the individual going by the strength of promises made. Many societies are increasingly abandoning their old members. With such mental dispositions the aged can look at what they can gain by being alone. They can use this time to develop skills that they were unable to develop when they were bringing up the children. Or they can start an old people's network, read and write, etc.

Companionship

In chapter one, we talked about a combination of effort, discipline, attitudes and strategies.

While making a special effort to be friendly and companionable to each other, some couples take each other for granted, assuming that since they are together, they do not need to make a special effort towards strengthening their love. For a successful wife, companionship means to live with him wherever he lives. She does not stay back in the village while he lives in the city or vice versa. (This happens in many parts of Africa). They married to give each other company, so why live apart? Couples that live separately tend to grow apart. They do not bond strongly because they do not have the time to come to understand each other fully.

When husband and wife live together, they usually make mutual adjustments towards each other. When they live apart, these adjustments are never made and each continues to hold on to the selfish habits they had developed when single. Out of these selfish habits stories usually start spreading that either of them is having an affair. We have heard of wives in the villages having secret lovers and men in the city keeping mistresses or visiting red-light districts.

This separation also exposes the couple to dangers of sexually transmitted diseases, especially AIDS. In some cases his mother demands that the young bride should stay at the village and keep her company. While it is a good thing to keep her company, this is not a unique exercise that cannot be done by a hireling. And again this was not the reason why her son got married. He got married to start a family with his wife and children.

This is not calling for the trashing of old folks; it is calling for the understanding of the spouses' roles.

So What Is Love?

The home is the first school of love. Very early on we learn to appreciate being cuddled by our mother, being cared for by our concerned fathers and being accompanied by our siblings and other family members. From home we learn to love and to be loved. Unfortunately, also from home we can learn to abuse love. Love can be abused through violence, rape, incest, hate, and abandonment. A successful wife helps her family to try to avoid all this.

Love has been described as affection, tender feeling, devotion, care, fondness and many others. The girl who has suddenly met her supposed Mr Right may think she is in love when her knees feel wobbly, her heart rate increases and she wants to be next to him all the time. She is even ready to be his slave. This phenomenon is called infatuation or passionate love and can be calamitous.

Infatuation is not sufficient to make one decide to get married. It lacks intimacy and commitment. A leading researcher identified three important components of love, which must prevail for marriage to be based on love. These components are: passion, intimacy and commitment. Other combinations are possible, but our concern here is the combination that is ideal for marriage.

Passionate love is temporary and fades with time. If a marriage were based on this type of love alone it would not survive the trials that come with disillusionment. The intimacy and commitment to the institution of marriage help the two individuals to overcome the storms when the 'tragic' flaw surface in both. A tragic flaw is that weakness of character that hurts people around a person.

To a successful wife, love means total self-giving, forgiving and surrender. She learned earlier that it is in giving where one receives. She gives herself totally to her husband and children. She is always there for them. This presence may not be physical. It implies an attitude that cannot be replaced by mere presence. It is a willingness to do what has to be done to make the family happy. To her it is a mental disposition. When the baby is unwell at night, mother has to get up and attend to its needs, possibly taking the baby to hospital at an ungodly hour. The Bible says: 'If I speak in human and angelic tongues but do not have love, I am a resounding gong or a clashing cymbal'. One need not be a Christian to understand this beautiful message. The symbols used by this writer illustrate clearly what a wife should not be. Cymbals and gongs have to be manipulated by somebody to be heard. Love, on the other hand takes the initiative and does something. Her love is therefore committed, intimate and passionate.

Think about the case of Jill and Jim (not their real names). When Jill's father gave her in marriage he told Jim: "I'm giving my daughter to you to be your queen, not your servant. You must treat her like a queen and no more. If you are looking for a cook or house-help you will not find that in my daughter." Three months into the marriage Jill was already most disappointed. She expected Jim to prepare breakfast and take it to her in bed before he left for work. Jim would have liked to do something for his wife in the morning, but his job was so demanding that he not only left early but also had to travel out of the country a lot.

Jill fell into self-pity and depression and within three months they were divorced. She could not live with someone who was not ready to wait on her. Clearly, Jill misinterpreted the message that came from her father and it is doubtful whether Jill's marriage was driven by true love on her part. Further she appeared not to have the independence of mind, discussed above, to detach herself from the ties that bound her to her parents.

Unlike Jill, a successful wife is able to detach herself from those crutches and start building her own home.

Abuse of Love

We said above that love may get abused through incest, violence, insults, drug dependency and certain mental conditions.

The question is: How does a successful wife protect herself and her family from such difficulties?

Does it mean that a wife who says no to abuse and molestation is not successful? Of course not. On the contrary, one who conceals the evil perpetrated by these acts cannot be a loving wife/mother/friend, and therefore a successful wife.

Whereas a woman may prepare adequately for marriage by going for pre-marriage counselling and taking medical tests, she may believe that the man she has married is not going to be abusive, or commit any of the above offences against love. But there are exceptions to the rule. The mishap can happen. It has happened before, it is happening now and it will happen in the future.

The reason is that human beings experience evil, both within and around themselves. Such unfortunate happenings may not be due to evil intentions. Sometimes one may be ill. Or one may suffer from a mental condition that may make him abusive. A successful wife rises up to the occasion. She realizes that she has to take the bull by the horns.

If her man has become abusive or deranged, this is a cancer in the family body. Only surgery can remove the malignancy. But cancer must be diagnosed early to get cured. A wife that waits 40 years before complaining about her wayward husband has wasted truly valuable time. 'You can't teach an old dog new tricks.'

So how do we expose this evil? The first step is to acknowledge that it is there and it is real. The second step is to make it very clear to him that it is not acceptable to any human being. This second stage is the trickiest, because it can mean the end of the relationship. It has to be done with tact and understanding. It may involve getting experts to deal with the problem depending on its gravity.

The third stage involves getting professional help. This help can come from a marriage counsellor or a psychiatrist.

Sometimes couples take their problems to their friends and relatives. While this approach might have worked in the past, it may not be practical today. I have noticed many people destroy their marriages by getting "help" and "advice" from friends and relatives.

Such people may give advice according to what worked for them, but the circumstances in your situation are different. What works for Jennifer may not work for Maria. Getting help for one who has a condition does not mean blowing the whistle on him. When a house catches fire, wise people shout for help. Some people shun counselling or psychiatric help because they have the misconceived ideas that such help is for the raving mad. But being raving mad is an illness, which should be given due attention.

Consider the case of Dina and Jacob. They met at a cafeteria where Dina had gone for a soft drink with her friend while waiting to watch a movie and to unwind before going home to prepare dinner and sleep. Dina had seen Jacob rise from a humble primary schoolboy to a leading accountant but they had never sat down to discuss anything as boy and girl. Each of them was too busy growing up to notice the other.

However this evening when they met in the cafeteria they were instantly attracted to each other. They realized that they had many things in common and it was nice "seeing" each other. Within a month they were married at their local church.

To their parents and friends Dina and Jacob were the perfect couple. They were both earning comfortable salaries and soon had plans for the mortgage of their first house and each had a good car.

However, after the birth of twins, Jacob started showing signs of occasional irritability and his mood swung between mania and depression. During a manic episode they were seated at table. He got up, and using the table knife and fork he would play some music by striking them against each other close to Dina's ears. He would then get the children to sing with him at the top of their voices. Next week he would become so depressed as to spend long hours in bed and not want to talk to anyone.

Initially Dina thought he was reacting to the presence of the children. She did not seek therapy immediately; she assumed all was going to be well. Unfortunately for her during one of Jacob's manic attacks he raped her. Later when she sought therapy it turned out that Jacob suffered from a mental condition referred to as mixed episodes of depression or bipolar disorder, in which depression and manic excitements alternate.

Conflict Resolution

Conflict in marriage does not entail the end of love. It acts like a spice, since after solving the issue love is strengthened. The language and approach that a couple uses for conflict resolution determines whether they will continue with the relationship or not. Quoting Gottman: "A certain amount of conflict is necessary to help couples weed out actions and ways of dealing with each other that can harm their marriage in the long run.... "Couples who do not resolve the conflicts that are inevitable in any relationship run the risk of getting mired in negative emotions and eventually breaking up."[6]

Most books on marriage point out that communication is vital for the good development of marriage. A successful wife helps her husband to embrace the spirit of effective communication.

Gottman's four crucial and tricky words for conflict resolution are: Complaints, Criticism, Contempt, and Anger. Understanding the power of each of these words determines whether we resolve our conflicts amicably or whether we hurt and injure the other person.

Complaints

Complaints state our dissatisfaction with something. How we utter these statements determines whether we hurt the other person or not. Our tone of voice, our facial expression, and our attitude towards the other person are very powerful messages. Emuna Brauerman warns against generalizing, when we disagree with the other person, by making such statements as: "You always do that", "You are never here when I need you" "Why can't you ever help me?" Etc.

Sweeping generalizations can be very damaging. No one is always wrong. No one always behaves negatively, and to put the accused party on the defensive obscures the issue at hand (Emuna).

Not all complaining is useful, warns Wood.

"Cross-complaining, in which you counter your spouse's complaint with one of your own and ignore what the other has said, obviously serves no useful purpose. Shouting off a long list of complaints all at once – Gottman's 'kitchen-sinking' – is not productive either."[7]

Voicing complaints positively requires a certain degree of composure. It calls for presence of mind by asking ourselves: What do I want to get out of the situation? To hurt my spouse or to make him understand my hurt? Do I want him to reform after understanding?

"Words are like bullets. Used improperly, they hurt a person permanently. The scars remain for life."[8]

Criticism

No one likes having one's mistakes criticised. Some people hate criticism so much that they can leave an institution because of it.

Like complaints, criticism can be positive or negative. Constructive criticism applies to marriage as to any other institution. If we criticize when angry, we may fail to use the right words and instead of attacking the problem attack the person. A successful wife avoids such a thing.

Somebody once said "Every criticism should be made into a sandwich with the bread of praise on the either side." When criticizing, as when complaining, we should avoid generalizations like: "You are always lazy", "You are always thinking of your mother," etc. These statements do not make the person reform. On the contrary they may stimulate the person into stubbornness. This is because our style of criticism hurts. Criticism differs from a complaint in that it heaps blame on the person.[9] A complaint is likely to begin with "I" but criticism begins with "YOU": "You always do that." Criticism goes beyond a feeling: it is an accusation.[10]

Positively used, criticism can correct unwanted behaviour. Used as a weapon, or too frequently, it can ruin relationships. A successful wife puts her criticism across positively. When the husband is late for dinner, or does not appear at all, she tries something like, "We missed you very much when we had dinner without you," instead of, "You are always late for dinner," or "You are never here for dinner."

To her daughter who does not make her bed, instead of "You never make your bed," she tries something like "Your room looks very neat and presentable when your bed is made, what happened today?"

Positive criticism is more likely to yield positive results than negative criticism does.

Dealing with criticism

We meet criticism in many areas of our lives. Couples criticise each other, as do parents, teachers, superiors at work, fellow workers and also in-laws. A speaker at a recent function compared criticism to a wave that will drown whoever fails to ride it. Many people who level criticism at us usually do not like the way we do things. Others do not know how to criticise without hurting. By being tactful, a successful wife wriggles out of such situations without causing confrontation. A confrontational approach fuels the situation.

Anger

"Anger is a passion for good or evil. It can be channelled and used not only for our mental, physical, emotional health and maturity, but also for the improvement of our intimate relationships."[11]

It can also be looked at as "an explosion of suspicion, greed or excessive power."[12]

We may well shout at other people, or be angry with them, when the problem is ourselves. At times we keep grudges, and vent them on our spouses when they make the slightest mistake. It is important to establish the cause of anger, deal with it and if possible seek professional help.

A successful wife helps all in the family to manage anger, for it not to result in catastrophic acts.

Contempt

Contempt is more serious and damaging than criticism, complaints or anger. A wife who allows anger to control her for a long time can become contemptuous.

Marriage counsellors say that when a couple becomes mutually contemptuous, their marriage may be heading for trouble.

"What distinguishes contempt from criticism is the intention to insult and psychologically abuse."[13]

"Contempt adds insults to criticism."

One can tell a contemptuous person by his or her choice of words. Contemptuous words are hostile and inflict pain.

Wood and Wood give the following guidelines for resolving conflicts:

i) Be gentle with complaints. Kindness works wonders when starting a complaint, and remember to avoid criticism.

ii) Don't get defensive when your spouse makes requests. Remain positive and comply willingly and as quickly as possible.

iii) Stop conflict before it gets out of hand. Don't let negative thoughts about your spouse grow into criticism or contempt. Do whatever it takes to put brakes on negativity. Recall and relish the happiest moments you and he have spent together.

b) RELATIONSHIPS

Husband

The husband is the wife's first relative in marriage. She leaves behind her mother and father and clings to him and they become one flesh. He therefore becomes the first and most important person in her life. Sometimes after the honeymoon, when disillusionment sets in, couples may begin to take each other for granted. This attitude may lead to misunderstandings. A husband who used to call his wife several times a day during courtship may settle back in his work and find no need to call her during a whole day. He might argue that after all she is already his and at his home. The wife, who might be very busy doing something for the home, may feel neglected and start brooding. Such a state of affairs does not help either of them.

They should be able to make time for romance and keep the fire of their love burning. A wise wife leads her husband into doing what is best for the family without trying to boss him. God has endowed her with the necessary skills to get her husband to listen to her. A loved and cherished husband performs better in his effort to feed the family than one who is nagged, defied and resented.

A wise wife is a good listener. She listens to his concerns and consoles him when the going gets tough out there. He confides in her and she trusts him completely.

Sex

A book on a successful wife would not be complete without a word on sex. It is the most intimate expression of love, but at the beginning can cause anxiety in either spouse. This anxiety is normal. It should however not blind them to the joy that comes with sex properly understood. There are many books on anatomy and sexual response that their counsellor may recommend. These manuals will however not replace the importance of carefully exploring and discovering what is best and acceptable to both. Each couple is unique in intimacy, and communication of their needs and expectations is crucial.

Just as the wife keeps her family always interested in her dinners because of trying new aromas, lasting interest in sex is brought about by experimenting with new styles frequently and a constant interest in each other says a writer on the joy of sex.

Apart from different styles, a leading gynaecologist advises that the tone and strength of the vaginal walls make the difference between being sexually in shape or out of shape.

She recommends kegel exercises, which involve the tightening and releasing of the muscles that control urine.

She recommends the following procedure: she should contract the vaginal muscles hard, as if trying to stop her urinary stream and hold for six seconds. This is done 10 to 15 times a day for about two to three months.

If either of the spouses has not received adequate training or rather information about sex during the pre-marriage counselling, sex may be unsatisfactory to both. In some communities young girls are taught that sex is improper and dirty when growing up. Instead of the mother and relatives telling her that it is holy but must wait until marriage, they confuse her. The fear of having a sixteen year old getting pregnant is real for parents but this fear should not lead parents into equipping this future mother with the wrong information.

Maggie and Joe met in college where Maggie was a student of law and Joe was completing his postgraduate studies in education. Theirs was the proverbial love at first sight and within a year they were married. Maggie grew up in an estate of well-to-do neighbours and her parents were extremely strict. They did not allow her to socialize with other girls for fear of bad influence. Her friends were her cousins and the children of her parents' friends.

Joe grew up in a rural set up and did not have much time with his parents. He had to work in people's houses to raise money not only for himself but also for his poor family. He studied hard and did well in school. When he started working, he was determined to give his children one day the life that his parents were not able to give him.

In marriage, Maggie did not change the image of sex that her mother and relatives had given her. To her it was something dirty and not worthy investing time in. According to her, it was also not meant for pleasure but for procreation only. Whenever Joe tried getting close to her for intimacy, she would reprimand him and tell him that he only thought about one thing. Joe actually did not think about sex much because he also did not have anyone teach him about it and did not live with his parents long enough to see how they related with each other sexually.

However, he desired Maggie and felt frustrated when she turned him down and told him off. He confided to a friend that he was on the verge of seeking divorce from Maggie due to her 'frigidity'. The friend helped him to seek therapy. During therapy Maggie was helped to get out of her misconception about sex and view it as pure and enjoyable and not a chore.

The physiological and psychological differences between the male and female sexuality must be understood by every couple so as to avoid the problems that arise as a result of misunderstandings.

Occasionally men complain of wives ignoring their needs while women complain of men being interested in sex more than anything else hence the expression: "You always think of sex" by women who have not understood their husbands' physiology.

Other Differences

A successful wife understands the basic differences between men and women and this knowledge helps her to understand her husband's moods and needs.

A marriage counsellor once remarked that during sex men behave 'as if they are riding a stolen bicycle which they must get rid of as quickly as possible before the owner catches up with them'.

While this counsellor was concerned about men who did not take time to please their wives sexually, he was pointing to lack of self-control in the sexual act.

Biologically men are different from women. Physiologically and psychologically they function differently.

According to Paul and Lori, there are mental, emotional and physical reasons why men feel a need for "regular release". Paul and Lori say that in men the seminal vesicles which are like two bladders fill up. The fulness makes the body signal the brain that release is needed. Women do not have this urge for regular release as such.

"Tender, loving touches coupled with verbal expressions of love arouse women more readily than visual stimulation."

- "Men are more aroused by what they see. Visual cues such as watching a woman undress are more likely to arouse them."

The writer of a best seller identifies the following differences between men and women:

- When they have problems, men prefer to "retreat" from other people, even from their own wives, to sort out their problems. At such a time when the man wants to be alone, the woman might think that he is ignoring her.
- When women have problems, they like to share them with their friends. A woman will always look for a friend to share a frustration either with her husband or with the children.
- Men like to be recognized as heads of families without getting challenged. For them respect and honour are the greatest ego boosters. Women on the other hand like to be cherished and loved.
- Men are impatient; you would be trying to tell him a story then he butts in impatiently…what happened. Women are patient and would like to hear all the details of the whole story.
- Men are not biologically endowed with organs for nurturing the young ones but women have this endowment.
- As Munroe puts it, women think in grids while men tend to be linear in their thinking. Additionally, women are more emotional; their brains are more wired to be so while men tend to be more logical.[14]

A successful wife understands her husband's need to be respected and honoured. She fulfils this need by constantly praising him, and reminding him of her love and respect for him. Because she understands his need for regular release, she works with him to keep his sexuality under control. She does not rebuke him and tell him, "You are always thinking of sex."

Lori has this note for wives, "When the man constantly feels the strong need for release, it is hard for him to focus on the other important aspects of intimacy.

A little analogy: When you skip a meal or two, you get very hungry. Pretty soon you stop caring much about what or where you eat, just as long as you can get something to eat! It is something like this for men in the area of sex. When the man constantly feels a strong need for release, it is very hard for him to focus on the other very important aspects of intimacy (he's too hungry). The good news is that a man who has the regular release he needs, finds it much easier to approach sex in a way that is physically and emotionally fulfilling to both partners. Also he will find it much easier to be emotionally and spiritually intimate in other areas of marriage."

Apart from sex; men are also different form women in that while women are concerned with trying to create harmony and beauty in the house and making everyone happy, Men are more concerned with activities like sports and competitive activities like trying to make their business successful.

Men do not remember dates and past events as women do. When they are out with new friends one may ask a man: when were you married? He does not even remember the year. The woman quickly mentions the exact date, place and time and may later blame her husband for not remembering this important date. The man gets shocked when his wife tries to make a fuss over it later in the evening.

In conclusion, one can look at the home as the man's castle where he is king and receives honour and respect from his wife who is his queen. As queen, the home is her palace and she is loved and cherished by her husband.

In her effort to respect and honour her king, a successful wife listens to him and avoids trying to give him advice.

CHAPTER 3

FAMILY AND WORK

Balancing Career and House Work

It would be most ideal for any woman to stay at home and look after the family. Looking after a family is the most important task if a community has to transmit life and values to the next generation. This is a responsibility of both parents, but who should do what in the home is a question of mutual agreement. Mothers are better endowed with the qualities to nurture their young than fathers, so naturally they take up that responsibility. The father then takes up a paid job or a business to earn for the family.

This is however not possible for most of us. One has to bring bread to the table one way or the other.

Balancing the bringing up of children with professional work gets tricky when one has to make sure not to neglect either. When a successful wife goes to work, she makes sure that she leaves her children with a maid whom she has trained in her family values and duties. Sometimes she may have to scale down climbing to the top of her career so as to avail herself to the family.

The future of mankind depends on what parents do with their children today. A mother who sacrifices the prestige of office work or a high public profile to look after her children does not lower her dignity by so doing. She does the most honourable thing because she ensures proper transmission of family values to her children and gives them the best foundation in life.

This does not mean that she should not pursue her career goals or that she should become destitute and completely dependent. It only means focusing on the family more and postponing some career decisions.

Both parents are supposed to be good role models for their children. A successful wife works hand in hand with her husband to ensure that the children get a good foundation in life and are happy. Raising children is a major topic and comes later in this work.

Most mothers today have to balance a career and housekeeping. It can be easy to focus on one and neglect the other. To avoid pursuing one at the expense of the other, one has to be efficient and orderly. This calls for proper planning and getting necessary help. A mother who wants to climb to the top of her career in education, politics, accounting or even in the building industry, may have to scale it down while the children are young, waiting to take it up more seriously when the children have left school.

A working mother can also organize for a leave of absence to look after her baby and resume work when the baby goes to school or can be left at a day care centre or with a responsible maid. A responsible maid is not always easy to come by but thank God there are institutions today giving formation to maids who can be a boon to working mothers.

A mother of small children can get self-employed and scale her business to the most practical level. While looking after the children the self-employed mother may not aim at becoming a household name overnight. She will however have the opportunity to enjoy her career and looking after her family at the same time. There are many opportunities for self-employment depending on one's studies. A writer of a bestseller, Sharon Maxwell Magnus gives the following list of ideas for making serious money from home:

- Running an agency
- Consultancy
- Counselling and Psychotherapy
- Beauty therapy
- Craftwork
- Running a B&B
- Direct selling and network marketing
- Book-Keeping
- Dressmaking
- Catering
- Image Consultant
- Child-minding
- Interior Design

- Making money with your computer
- Mail order
- Complementary therapy
- Organizer
- Public relations consultant
- Teaching
- Freelance writer.

Relatives

Both husband and wife know that their relatives are as important as their friends, and that they cannot choose to keep these folks away completely. The list of relatives, his and hers, can be long. The couple has to try and maintain a good relationship with the relatives as with friends. The new couple, however, must not become dependent on relatives. Wife and husband should depend on each other, but independent from friends and relatives. A young wife might try to call her mother or aunt every time she has a minor problem. She should learn to solve such problems with the help of her husband and relevant professionals.

A good relationship with relatives helps the children to learn about public relations early in life and from home. These relatives can be the first friends that the children notice in the home. How a successful wife handles her relatives helps her husband to accept them. If she presents them to him as people who are worthy of respect and friendship, they will receive his friendship. If on the other hand she shows him that they are not worthy of respect, then he will not respect them. The same is true about the husband's relatives. If he convinces his wife that they are not worthy of respect they will find a very indifferent sister-in-law.

Grandparents are very important to growing children. They offer a sober, mature and unconditional love to the children. They are a testimony of life to the children.

> Grandparents are important in every family. They are the guarantors of affection and tenderness, which every human being needs to give and receive.
>
> They offer the little ones the perspective of time, and they are the memory and richness of families. In no way should they ever be excluded from the family circle.
>
> They are a treasure, which the young generation should not be denied, especially when they bear witness to their faith at the approach of death. (Pope Benedict XVI, *Address to the Fifth World meeting of Families,* Valencia, 8 July 2006).

The way a wife treats the grandparents of her children is a great message and lesson to her children that "this is the way to treat the aged."

When one of them is sick and ailing, he/she requires great care from the family and can add a strain to the already stressed wife. This however becomes a good opportunity for her to teach the young ones about love and service. She can demonstrate to her young that to be young is not to be inferior and serving is not an inferior activity either. I am saying this because of those of us who feel that serving one who is infirm is inferior and should be done by a hireling.

In-laws

Having problems with the in-laws is a pointer to more serious problems in the marriage itself. A properly bonded couple who know the importance of the new family unit they have created will be able to deal with the problems of the in-laws very well. This will be so because they work as a team. It is the responsibility of a successful wife to help the husband to make the adjustment.

A successful wife knows that she has to handle in-laws tactfully. In-laws can be a constant headache or a source of joy and support depending on how one handles them. Most newly married women complain of the mother-in-law being bossy and issuing unreasonable orders around the house. This is most likely to happen when the new couple did not make adequate preparation to get their own house and they therefore have to stay in the house of one of their parents.

In some African homes a newly married woman suffers great stress from the siblings of the husband because they have been conditioned to depend on their elder brother in the case of a deceased father or a father who is a dependent himself. (The man suffers a similar crisis from the relatives of the woman)

In-laws are blood relations of the new couple. Because of the strong bond between siblings it may happen that the new wife is perceived as a threat to this bond by them. His detachment from his siblings requires great tact because if handled carelessly it can lead to pain and hostility.

The Mother-in-Law

The relationship between a mother-in-law and a new bride can be a nightmare. Some mothers-in-law fail to accept the new bride and perceive her as a threat to the love they enjoyed from their sons. This usually happens when the mother has had a dysfunctional marriage of her own, and has switched the love she felt for her husband onto her son. She lives for her son and he is the whole world to her. The new bride then finds herself in a love triangle and matters here can get complicated. The new bride can also be a nightmare to her mother-in-law when she does not allow the old woman access to her son.

A successful wife knows that 'trying to complain about her mother-in-law to her husband will only serve to make a puppet of him'. A successful wife talks to her mother-in-law tactfully, explaining to her that her behaviour and demands may not be acceptable to her as a new bride. If they are to live together it must be on a mutually agreed solution. Then they agree to work together as a team and not as rivals. It is very important to make it clear to the mother-in-law that the relationship between her son and his wife is different from the one between her and her son. Therefore the two roles should never be confused.

A successful wife uses tact to deal with her mother-in-law amicably.

In some parts of the world the mother in-law is referred to as the devil, in others as an angel.

Happy is she who finds a loving and supportive mother-in-law!

Woe unto you if you don't.

In a patriarchal society, where the new bride has to live with the parents of her husband, the extended family life can be very difficult for her. She is expected to play 'maid' to all the family members and some families can be very large.

The new bride must understand that the most important family is the new family unit, that she and her husband got together to start it and should be safeguarded by all means.

A successful wife also knows that her home is a school. Her children are watching when growing up. The treatment that she gives to her in-laws is the treatment she expects her *children to give when they become sons and daughters-in-law.*

The Father-in-law

Fathers-in-law become a problem when they start making unreasonable demands on their children or even start making passes at their daughters-in-law.

The most humiliating experience for the newly married girl can be to wake up one day and get accosted by her father-in-law in a corner of her own house or even in the house of her in-laws.

It is not always the mother-in-law who can inconvenience her children. The father in-law can also be a bother.

The Sister-in-law

A sister-in-law can be a great friend and a source of support and joy when handled with tact and understanding. The sister-in-law who has not been helped to understand that her brother has started a new family unit of his own can continue clinging on to him and become a real rival to the wife. Rivalry in these relationships can cause great pain to the parties involved if it is not dealt with from the beginning.

A couple can help a sister-in-law by getting her an occupation to keep her busy and discourage her from sinking into self-pity and depression. A sister-in-law can either be her sister or his sister.

Although husband and wife should handle their siblings as a team, care should be taken to avoid isolating and alienating them.

A successful wife makes it clear to her sisters-in-law that she has expectations from them just as they have expectations from her. However all the expectations and actions must be tailored towards the protection and welfare of the new family unit that she has established with her husband.

It can be painful to let go of a brother just as it is painful for a mother to let go of her son. A successful wife helps them to make this adjustment painlessly.

A successful wife helps her husband's sister to understand that she needs time to learn to live with her husband and that this time brooks no interruption and should be respected. Most sisters-in-law understand and respect this. Few, however, may not respect this out of ignorance, which requires a firm approach.

Mariettah got married to Joe, who was an only son among eight siblings. Joe was also the firstborn of the eight. When his parents retired from formal employment, Joe took over the responsibility of running their household. Although three of Joe's sisters were already working and getting good salaries, they were not as responsible as he.

Four of Joe's sisters moved into Mariettah's house to live with her and continue getting support from their brother. Initially Mariettah did not mind this since she expected them to be friendly and supportive.

But the girls were extravagant. They criticised Mariettah's cooking and her style of doing things generally. They also got into her wardrobe and put on her clothes and shoes whenever going out. When she told them that she did not like it they told her that their brother had bought those items and they had a greater right to them than she did.

Occasionally their mother would visit and the five would expect Mariettah to wait on them and serve them like queens. Mariettah felt invaded and unable to take it any more.

She sought therapy and is now living happily with Joe and their three children, while still being great friends with Joe's sisters.

Brothers-in-law

Like sisters-in-law, brothers-in-law can also suffer sibling rivalry and cause tension in the lives of the new couple. Brothers who were used to going out with their sibling find it hard to let go, especially if the newly married brother was their source of financial support. A successful wife handles them with tact and seeks therapy when push comes to shove. She does not go to her mother or mother-in-law for mediation.

Chapter 4

CHILDREN AND PARENTING

Parents know the joy that every child brings into the home.

It is with the arrival of the children that a couple feels fulfilled. Whereas it is joy to some people, to others it can mean trouble. Jealousy, the green-eyed monster as some call it, can rear its ugly head. The man can easily feel that the children have replaced him in his wife's heart. A successful wife assures her husband that his place cannot be taken by the children. They should work together as a team to bring up their children in the best possible way.

Although the arrival of the children is such a great joy, bringing them up can be a great challenge. It is however insuperable only to those not prepared for it. Children require proper training at home to become useful members of society. It is also at home where they should get introduced to the idea of professionalism.

Three methods of parenting have been identified by Diane Baumrind:

The first method is *permissive*. Parents are warm and supportive but make few rules or demands and usually do not enforce those that they make. Children brought up this way do not become responsible. They lack self-control and they do not respect authority. Children who cannot respect authority have trouble fitting in institutions outside the home because law and order are a must in such institutions. Many of us have observed children who get noisy and disruptive to visitors while the mother does nothing about it. Some mothers will not ask the children to give them space to talk to visitors.

The second method is *authoritarian*. The parents make the rules and expect unquestioned obedience. They use such language as: you must do this. If the children question why, the parent's response is usually 'because I say so'.

Children are not taught to think and weigh out things for themselves. They are only expected to take orders and produce results like robots.

Such children face serious problems when they leave their parents' home. They are not able to make firm decisions for themselves and they grow up withdrawn, anxious and unhappy.

The third method is *authoritative* (as opposed to authoritarian) parents. They set high but reasonable standards that can be attained by their children. They discuss these rules with their children and help them understand and internalise them. This is done with warmth and love. Here children are trained to think things out for themselves. The children of authoritative parents are mature, self-reliant, self-controlled, assertive, socially competent and responsible.[15]

It is clear that the authoritative type of parenting produces the best-adjusted children.

In training her children to be responsible adults in future, a successful wife focuses on giving her children the training that best prepares them for their careers. This can start small, in things like asking them to make their beds, return their plates after dinner, reading/singing to the family after dinner, put order in their wardrobe and polish their shoes, enabling them to perform more difficult tasks. A successful wife teaches her children public relations by allowing them to interact with their grand parents and other members of the extended family.

Children also need to be brought into contact with animals so as to strengthen their immune systems.

The Permissive Parents

We are seated in the headmaster's office with John and both of his parents. They have reported back with him after a two-week suspension for use of vulgar language while talking back to the teacher. We are members of the disciplinary committee who have been called in to look into John's case.

John's parents blame the teachers for not instilling discipline in their students and feel that in other schools such cases do not occur. The principal has been in education for a long time, and notices children from permissive parents the first time he talks to them in the office.

They show a clear lack of respect for authority and do not exercise self-control.

When invited in the headmaster's office, John talked to him with hands in his pocket, reached for items on the headmaster's table and made uncalled-for comments like: "Why don't you ask my father for a better letter opener." This team recommended sessions with the school counsellor.

During therapy it emerged that John had received firm directives from his parents. They would ask him to make his bed but would not insist if he did not make it.

John also had a problem with overeating.

The counsellor put John through a series of exercises to develop self-control and respect for authority.

The Authoritarian Parents

Mark is a forty-year-old company C.E.O. He did not climb to this position by chance. He is intelligent and capable. All his school reports showed that he was intelligent and able to organize ideas.

However, Mark has one major weakness. He takes a very long time to arrive at a decision and many times his assistants have to fight very hard to come to an agreement with him to implement it. Fortunately his deputy understands him well and knows how to get him into agreeing with them.

Mark also finds it very hard to trust anyone. He is always suspicious that some assistants come up with ideas actually snatched from him. He takes each one of them to task to show that they have not turned into competitors.

His wife also has problems. When she serves him a pie, she has to share some of it with him, and not out of love. She has to prove that she has not laced it with something undesirable.

Mark's problems started when he was a small child. His father was a heavyset man with thick eyebrows and moustache who had grown up as an orphan and was an introvert. He never discussed anything with anyone and his word was law. Not even his wife could question what he said.

When Mark was asked to do something by his father and asked why, the answer was "because I say so." His mother would tell him, "If your father has said it, then it must be done."

Security for Children

Immunity

For children living in the urban areas a trip to the country to see grandma's goats and chickens can be very helpful. One scientist argued 'We do not develop immunity against pathogens that we do not come across'. Probably this explains why children instinctively eat soil and stuff dirt in their mouths.

While providing immunity to the children by exposing them to domestic animals, a successful wife knows that it is important to protect them against predators in the form of child abusers.

Security

Child abuses can and take place even in the sanctuary of the home, where children have learned to relax and expect the best.

While child defilers waylay children on their way home from school, or while playing outside, some abuse takes place at home.

People that the parents least expect to be child molesters perpetrate this crime.

A friend or relative who visits and has to spend the night or week in your house might be the same one who abuses your child when you are away.

There are also unfortunate cases where the father defiles his daughter and begs the wife or even the daughter to be quiet about it so as to safeguard "the integrity of the home."

A home where the father defiles the daughter has no integrity. This act should be exposed to therapy at least for the sake of the child. The bigger or older siblings may also defile the younger ones. There are also reported cases of mothers defiling their own daughters but those are rare.

A child should be trained from the earliest possible time to say no to alluring things like sweets and other presents from strangers. They should also scream when somebody tries to touch them in areas not permitted. They should report any uncomfortable touch by anyone. If possible parents should organize the transport to school and back so that the child does not have to walk a long distance to the house after getting off the bus. If possible there should be a grown up to wait for the child.

Those who walk to and from school should always do so in groups. A child should not walk alone.

They should get enough to eat so that the allure of food like chips from strangers may not entice them.

When alone in the house, children should know that they should not open for any strangers unless so instructed by their parents or guardian. While left on their own children should not tell strangers on the phone that they are on their own.

If parents maintain an open dialogue with their children, some of these mishaps can be avoided. Children will always report any unacceptable behaviour to their parents. It is very important for parents to learn to listen to their children attentively and always try to understand what the child is trying to say. They may be reporting something very serious.

Education

A successful wife knows that the influence education will have on her children is as strong as her influence on them. This is why she chooses schools that offer the set of values that the family espouses. The choice of the school will also depend on the distance from the school.

A writer in social psychology says, "Of all factors involved in attitude formation, education consistently stands out. It has as strong an influence on the individual as the parental, political orientation and religious affiliations."[16]

School is not the only place where children learn. They learn from home whatever Mum and Dad teach them, they learn from Sunday school, from a visit to the Animal Orphanage and even during those school tours to farms, museums, wineries and the rest. They also learn by helping Mum make a cake in the kitchen, washing dishes, preparing beans for boiling etc.

Play has been identified as one of the exercises that help children to learn. Play should be organized so as not to deteriorate into a noise-making exercise.

Attitude Formation

A successful wife helps her children to form positive attitudes towards other people and animals. As a mother, she knows that she shapes the opinions/attitudes of her children, from before adolescence, by what she approves of, what she does and what she says. It is reported that many teachers in the pre-primary schools say children always report the opinions of their parents. For example, when a teacher tries to teach a child to pray, this child may say; “My mommy says that we make the sign of the cross when we pray”. To most children in this age group what Mum or Dad says is gospel truth that shapes their world. To many parents this is not a very difficult age because the children trust their parents and teachers completely.

As teenage approaches, boys and girls begin to trust the opinions/influences of their peers, television, relatives and friends. They begin to question the authority of the school, of the parents and of the church. Sometimes teenagers decide to disobey despite having parents who are authoritative.

The good news for parents is that this phase soon ends when they remain firm in their expectations. Adolescents are said to ‘freeze’ the good attitudes that the parents inculcated in them so that as adults they may not change much despite peer pressure.

Our intentions are not visible, but the people we encounter daily know whether we love or hate them without a word from us. They read our minds and our thoughts from nuances of expressions on our faces. The words we use to describe our ideas and our thoughts also communicate our intentions. In other words if we hate people we do not need to say so openly. Even the most cunning who try to pretend to love by smiling artificially soon twitch a muscle somewhere that says something different from what they say.

A successful wife’s attitude remains positive towards her children and the family despite the ups and downs of life. She goes about life with the attitude "I want to serve". "I want to love, I want to be there for others, I want to motivate others." An attitude is not something one feels one day and forgets the next. Attitudes are enduring and to a successful wife they are positive.

Social psychologists suggest, "When we have a negative attitude towards something, we also have unfavourable beliefs about it". For example if a wife believes that "my child is useless" or "does not work hard enough," this belief will affect her emotions. She begins to dislike her child and probably regard him or her with less respect than deserved.

Motivating Children

Children are born into a home. Home is the first school that the child experiences. How the children will perform first in school and later in life depends largely on the foundation they get at home.

Any teacher will tell you that the children who love school and learning had the foundations laid at home.

Whereas the teacher has a duty to motivate children to learn and do well, parents have an obligation to create the attitudes that foster this motivation.

It can be very difficult for a teacher to motivate a child who has developed an indifferent or negative attitude at home.

Why do we need to motivate children apart from, or against anxiety, self esteem, sibling rivalry, laziness etc.? Psychologists have identified motivation as a crucial factor in enhancing learning.

What is motivation? Annetta Miller in her article 'Lighting a fire' says, "if confidence is an internal voice that says '*I can do it*', 'motivation is the voice that says '*I want to do it*'".

All children are born with a certain amount of motivation. This can be noticed from an early age when a toddler pulls a tablecloth off the table and brings down all the glasses or when it picks things from the ground and puts them in the mouth or ear.

Parents can do a lot to kindle this fire, to dim it or even to put it out altogether.

How parents react to children's behaviour determines whether it will be repeated or not. If a child's curiosity or behaviour (for example pulling the table cloth) is met with disapproval, (which may include punishment) it may never be repeated and sometimes the child may lose interest in other things and even get withdrawn. If on the other hand this behaviour or rather a child's curiosity is met with approval, the action may get repeated.

"The parent's reaction determines how the child feels about this learning process and can mean the difference between a motivated student and an indifferent one." Adele Faber, author of *How to talk so kids can learn* says, "To reach a child's mind, speak to his heart".

Annetta suggests that parents should bottle up this motivation in small children and save it for school days when children find many reasons for discouragement from their peers and the environment.

Students learn different subjects. The motivation one of them may have for arithmetic may not be the same as that felt for geography. So how do we equalise the enthusiasm for number work with that for geography?

There is no simple answer to this but the sum total of parental attitudes towards the child will determine whether the child gets motivated or withdrawn.

Getting Kids to Solve Problems Themselves

Getting them to solve problems themselves is the way out of a motivational conundrum. Says Annetta: "Children can use reference materials designed for their levels to solve problems that they encounter during their homework." Parents should not provide them with ready answers.

Although the relevance of homework in learning has been variously questioned, parents should encourage their children to do it, as it gives the children practice in what they learn in school. When parents take interest in their children's homework, the children realize that it is important and their own interest is stimulated.

Being Available

Children feel more confident when their parents are around. They find homework easier to handle when

Mum or Dad are around and ready to answer questions. The company of the parent alone gives children the assurance that what they are doing is important.

Discussions

Many times it may not be possible to discuss all subjects in detail with your child since no parent can be in a position to discuss subjects with the child the way a teacher would. But asking questions on the subject at hand helps the child to focus on the subject better.

Encouraging them to read stories and retell them helps towards giving the children language structure and vocabulary.

Friendship

It has been said that whoever likes to have friends must be a friend first. It is important for parents to cultivate friendship with their children because friendship is a universal good. Talking nicely to the children makes this friendship grow. When children are loved they are treated as friends, and this boosts their ego. They perform better in school as much as in other social activities.

Parents can teach their children "positive habits of mind" – attitudes and ways of thinking that lead to achievement and well-being.

It is hard for children to be motivated in school if they do not feel good about themselves.

Raising Their Self Esteem

Psychologists have established that children who are motivated to achieve positive things in life have a high self-esteem. Parents' style of doing things is crucial in enhancing the self-esteem of children. If they are treated as equals they feel recognized and appreciated.

Love also helps to boost the children's ego. Unloved children tend to perform poorly even with a high level of intelligence.

Another factor that helps boost the child's ego is the attention received at home long before school. Attention means taking interest in the child's activities, whereas love is shown through affection. Both however make the child feel that he/she is important.

Containing Rivalry

In a family, siblings tend to vie for the attention of the parents.

A successful wife works with her children to overcome this from the time she gets her second child, and does not allow sibling rivalry to degenerate into hatred and violence.

She shows her first child that the second one is not more important than the first is, and that they will work as a team to help the new baby to grow and be strong like them. When the older child realizes that he/she with Mum and Dad form a team helping the new baby, the firstborn loses any hostility he might have felt for the new baby, becoming cooperative instead towards supporting and protecting the new baby from then on.

This bonding of the new baby to the older sibling should start early, when the mother is expecting. She should prepare the older sibling to wait for the new baby with positive enthusiasm. She does this by saying positive things that will happen to the older sibling because of having a brother or sister.

There is another form of rivalry that may mar the relationship between man and wife: professional rivalry. Many a man believes that as the head of the family he should earn more than his wife. But this not always happens. Often the wife earns a higher salary than the husband, or she may even be the sole breadwinner. If not handled with caution, this situation can cause serious tension.

A successful wife handles such a situation with tact and understanding. She does not belittle her husband because he earns less. Their marriage does not revolve around finance alone. Their love is not based on finance or other material things. She makes sure that he feels respected as the head of the family.

Chapter 5

FRIENDS

The newly married couple makes new friends while keeping the old ones. The old saying that we should make new friends but keep the old, one is silver and the other is gold also applies to them. The wife's friends become her husband's friends and vice versa. Each of them must ensure that the time they spend out with friends is reduced, so that they have enough time to be together.

Friends are important to individuals as well as to families. They support us in times of difficulties, they share our joys, but they can also distress and even betray us.

We all need friends. They can be a great source of joy and support. Somebody once said that to be successful one has to be able to make friends from all races and ages. A tall order, isn't it? Friendship is a universal good and transcends race and creed. But when one has been betrayed by a friend, the tendency is to withdraw like a hermit and forget all.

Like relatives, friends need to be treated with understanding and care. They may at times wrong us, for to err is human, but they will also be there when we need them. Let us also remember that we also may wrong them and they forgive us. If we understand the joy of forgiveness we should extend it to our friends.

Nice remarks to friends always make them feel good.

Chapter 6

WHEN THINGS GO WRONG

Betrayal

No one expects their loved ones to have an affair behind their back. When it happens, reactions can be as varied as the people affected. Some people feel the same sense of grief as that caused by bereavement. Others feel ashamed of being associated with such a person and others get so enraged as to feel able to kill. Others yet experience depression and have to get professional help.

At such a time, a wife may wonder whether she will stay put, or pack up and go, or ask him to leave. A betrayed husband may feel the same.

Dealing with betrayal is the most difficult challenge in marriage. It is most important to understand that the problem is the betrayer, not the betrayed.

Understanding the weakness that leads to such behaviour and helping the offender to sort it out goes a long way towards healing the hurt. Healing after betrayal does not happen instantly. It keeps surfacing in the psyche of the offended party, giving a bad feeling. It needs time to heal, just like a physical wound.

A man may betray his wife by having an affair or by paying a prostitute because he feels that his wife neglects him or does not understand his need for constant release. Other men betray their wives out of carelessness and lack of self-control. A talk with a counsellor or a good friend helps the betrayed party to recover and the offending one to practise self-control.

Drunkenness

Spouses who fall into the trap of alcohol do not drink out of choice but out of compulsion. They become unable to disengage from the drink. When they are not drunk they feel sick to the bone.

Most people experiment with alcohol when they are young, usually because of peer pressure. Those counselled at home about the dangers of alcohol and drugs may not get addicted and become victims. Others take alcohol to deaden feelings of pain and trauma. A child whose father or mother suddenly leaves home may start drinking to dull the feeling of rejection and loneliness. Talking to such people may not help them stop the habit. The non-drinking spouse may require professional help, which is not the same as blowing the whistle on the drinking husband or wife, or as giving up. It is sign of strength.

Alcoholism should be treated by a medical doctor, because it is a disease. The alcohol itself can lead to serious liver damage, brain damage, pancreatic cancer etc.

Violence

Alcoholism can lead to violence. A violent and abusive spouse is no longer the loving person one married. Violence can make a marriage miserable, and should never be allowed to develop. A man does not wake up violent one good day. Violent attitudes grow slowly, and should be identified early. A discussion with him before marriage can tell whether he is likely to become violent or not. She should ask about the methods his parents used to punish bad behaviour, or what he would do when provoked into anger.

In some communities violence is used as a means of solving disputes and disciplining children. Some Christians misinterpret Scripture and literally whip their children for the slightest misdemeanour. The saying "spare the rod and spoil the child" (Prv 13:24) calls on parents to instil discipline in their children. Beating of children is done in most instances by elders who want to gratify their passions and who were probably abusively punished as children. There are very many incentives to good behaviour available to a loving parent without having to resort to violence.

As there are deterrents to bad behaviour available to loving parents, if they resort to violence it is out of carelessness or impatience. Violence not always stops unwanted behaviour. It has been said that the victim of violence smoulders in resentment and humiliation. Like a boxer, he either recovers and returns to the fight wiser for the experience, or retires from the ring angry at the unevenness of the bout.

A wife whose husband turns violent should seek professional help immediately and make it very clear to him that violence is not acceptable.

The table below gives a list of deterrents to bad behaviour and incentives to good behaviour available to teachers and parents in lieu of violence.

Incentives to good behaviour	**Deterrents to bad behaviour**
Quiet appreciation by parent or teacher	Deprivation of privileges
Rewards	Firm insistence by the parent /teacher
Praise	Isolation
Elevation to positions of leadership	

Money

Money in marriage can be a major cause of friction and even divorce. Each of the spouses usually has a different spending/saving pattern because of their diverse backgrounds. They should discuss these and come to an agreement on their spending and saving.

They should always remind each other about cutting their coat according to their cloth and avoid keeping up with the Joneses.

Sickness

"In sickness and health till death do us part." These words should echo in the hearts of the spouses so long as they live. One of the functions of the family is to provide security to its members in times of sickness and other calamities. If a spouse becomes ill it is the responsibility of the healthy one to make sure that the other is well looked after, and if the illness is terminal, that he or she die in dignity.

If the husband is the sick one, the wife has to get means to generate income for the family. To do this she will have to leave her husband in the care of a house help or an unemployed relative who can do the work happily. Arrangements should be made to ensure that such a relative is properly remunerated to avoid tension.

If the wife is ill, arrangements should be made to get someone to look after the children for the duration of the sickness.

When There Are No Children

Given that the primary end of marriage is the procreation and upbringing of children, then naturally a childless couple cannot help but feel that something is not right with their union. Sometimes no matter how much a married couple may want children, none comes. If they are not informed, they may start blaming each other for their lack of children.

A couple that fails to get children after some years of marriage should seek the services of a doctor, with the view of ruling out the possibility of sickness. Whatever the medical findings, they should be able to accept it as God's will and learn to serve him in that state.

Sometimes it is not the childless couple that is disturbed by this state; their community can put undue pressure on the man or woman to look for a child extra-maritally. The parents of the childless couple may start demanding grandchildren and insisting that their son or daughter should marry someone able to produce children.

The couple should not bend to such pressure and should ask God to focus their energies on activities that can divert their minds from thinking about children all the time. After all when children grow up, one good day they will go, and eventually the couple is left alone. One should not overlook the fact that children are a gift from God.

While parents invest in children, the childless couple can invest in projects that can take care of them in their old age. Such projects can also form part of their apostolic mission towards serving God and their fellow men.

History portrays many married but childless men and women who gave their lives completely to God and are now saints. One such woman is Victoria Rasoamanarivo of Madagascar.

BLESSED VICTORIA RASOAMANARIVO

Victoria was born in Madagascar in 1848 when the Christian faith was under persecution. She became a Christian at the age of 15 and wanted to be a nun, but instead she was given in marriage to a non-Christian officer of the royal army. Victoria was always faithful to her husband even though he led an immoral life. He was critically injured in a serious accident in 1888, and before he died she baptised him. Victoria, now 40 and childless, decided to fulfil her dream and devote herself to charity and the care of the sick. When the missionaries left on account of the persecutions, she kept the Christian community strong in the faith and even pleaded with the authorities for permission to use the churches that had been closed down. When the missionaries returned they found a thriving church thanks to Victoria. She died in 1894. Pope John Paul II has called her "a true missionary" and "a model for today's laity".[17]

CHAPTER 7

HINTS TO MAKE A CHEERFUL HOME

A home is not made cheerful by letting in the sunlight through open windows. However beautiful the light of the sun is, it does not match the smiling faces of a happy family.

A successful wife goes about her activities happily, and tries to ensure that her happiness spreads to the family. She talks to her husband in a cheerful and loving manner. She knows that her moods are contagious, and that she can spread gloom as much as happiness. How does she achieve happiness? She has dispositions that help her to stay calm and composed even during stormy times.

Before Children Arrive

Before children begin to arrive, it is the great opportunity for the couple to adjust to each other and greatly enjoy the pleasure that God has set aside for that period. Their time is not yet stretched by the demands of a family with children and therefore they can spoil each other to the maximum.

That same time is also a time to get to know each other's friends and relatives, thereby enlarging the circle of their acquaintances. She can take advantage of this time to entertain them with ease. And she can also develop her career during the same time.

Keep up with the Joneses? Never. Not a successful wife. She cuts her coat according to her cloth and knows that the neighbours' situation is different and may not last that way. No two families are the same, and each can find happiness even with meagre resources. It is not material things that bring happiness to a home, however much they may mean to a given member of the family.

In an effort to be like their friends or relatives a couple may move into an expensive house that they cannot afford. They can also buy cars so expensive to run as to compromise the quality of their children's education or even the quality of their meals.

A successful wife discourages her husband from this kind of spending and helps him to focus on the right priorities.

A successful wife knows that if she cleans every nook and cranny of her house she will not need to do so tomorrow.

She knows that if something has to be done it is better to do it today. Procrastination burdens the mind and fogs the attention on the job. Paying attention to detail pays in more ways than one.

Before children arrive she has all the time in the world just for the two of them. This is also the period when she enjoys those candlelight dinners and takes advantage of the absence of other people to dress for him!

A Successful Wife Keeps the Flame of Love Burning

To be cheerful, the fire of love must be kept burning like the rich embers of a fireplace that are kept glowing by the addition of more firewood. A successful wife does this by little acts of love to her people. A tender touch to her husband, a kiss to baby and an attentive look at her son's homework are all means to this end.

A successful wife also shows courage in difficult times. She defies harmful peer and social pressure to stand for what she believes. When she loves her family and her husband she does not give up on them because of problems. It may be fashionable in some communities to boast of being in one's third or fourth marriage. A successful wife does not bend to such pressure. She knows that all men and women are influenced by emotions and conflicts originating in the community where one grows up. A successful wife also knows that one can spend a life time vainly hopping from one marriage to another.

Through the practice of this type of courage, which is an aspect of fortitude, a successful wife can become a good public relations practitioner. Sometimes people who talk a lot and have abrasive manners put us off.

They even lack tact and can be a cause of frustration. Such people may be friends, relatives or even strangers. This type of courage gives a successful wife the strength to listen to these people and sift the grain of information from the chaff of gossip.

At Home with the Children

A successful wife is proud to serve her family and to make her family members happy. At table she initiates conversation over meals to make her family members cheerful. She shares her son's fun with language in school, her daughter's fun with numbers and Dad's day in the office. Her meals need not be elaborate but they are fun. To her meal time is a special moment for all. She also welcomes friends and relatives into the house and greets them warmly.

She insists that all should be in for breakfast and dinner, thus instilling discipline in her children. The children are not allowed to leave table before everyone has finished. The one who tries to swallow food hurriedly so as to rush out and play has to sit there and watch the others eat.

ANNA

Anna is always happy to have her family together for meals. She serves her meals in an extraordinary style. She may cook the same staple daily but in different styles. If she has only rice for the whole week, today she makes fried rice, tomorrow pilau, the day after rice and beans until the week is over with a different recipe for each day. But it is not only the quality of her cooking that keeps her family present for every meal. It is her charm. She makes mealtime such a warm time that everybody soon engages into conversation and laughter. Her husband and children are always looking forward to getting home and relaxing. She does not go round telling her husband and children how much she loves them; she shows it in deeds. They see love and feel it in every corner of the house and in the countenance of their mother and wife. Her husband is at peace, and therefore able to focus on his work and earn money for the family.

NANA

Nana's home is spotlessly clean. She studied Art and Home Economics at the university and practises what she learned to the letter. Her house is more of a showroom than a home. She experiences great pain asking a friend to come home for lunch or dinner, because this might interfere with her order of things in the house. She does not allow anyone to step on her carpet with shoes on, and therefore members of the family always avoid that area. Most of them feel humiliated by being asked to remove their shoes on entering the house. She finds her husband and children a great bother, and remains cold towards them.

Nana's children feel more relaxed and at ease with the housemaid than with their mother. When a child tries to get close Nana reminds them that the maid is there to mind children's little worries. Nana is very careful about her makeup and feels that her children may interfere with it. Her husband is very happy to be away 'making money' and therefore unable to come home early any day of the week.

To maintain a cheerful home, the wife has to be cheerful herself. She cannot give what she herself lacks. She believes very strongly that nothing can happen to anyone else unless it happens to her first. This is in the sense of not being able to give what you do not have. She therefore takes practical steps towards remaining cheerful.

Health

A successful wife nourishes her family on non highly-processed foods. One should guard against obesity by avoiding junk foods.

A successful wife knows how to be the first family doctor. She does not have to study anatomy and physiology to do this. There are simple but important things she does at home that reduce the number of times one of the family visits the doctor. These include:

Washing hands

This simple exercise gets rid of a lot of dirt picked up from matatus, buses, shaking hands, visiting the toilet, working in the garden and holding door knobs and handles. It is a health investment, because it saves the money we would spend in the hospital treating avoidable diseases.

One washes hands not just by rinsing the fingers and drying them with the towel. One needs to use soap and rinse the hands thoroughly at least three times. This way even if children use their hands to handle food, they will be safe.

Sleep

To be cheerful, a successful wife makes sure that she gets a good night's sleep. A body that has not rested properly at night is neither cheerful nor productive. To get enough sleep she goes to bed and wakes up at fixed times.

"Get enough sleep every night." This prescription is as old as mankind itself yet it is not possible for most people. Life's stresses and the struggle for survival have made it difficult for the average person to sleep more than six hours a night. They may spend long wakeful hours in bed but that may not be adequate.

Sleep manuals recommend that one should sleep for eight hours a day.

Experts say that our mental conditions determine whether we sleep like a baby or toss and turn\throughout the night.

We all would like to sleep like babies but our lifestyles do not allow it. High levels of stress and anxiety prevent many from sleeping.

Dr Paul Clayton recommends steps that can help us sleep better:

1. Take more exercise during the day. But if your routine is such that you cannot exercise until evening, keep it gentle.
2. If possible get some exposure to daylight at least once a day.
3. No naps during the day no matter how drowsy you may feel and no coffee or tea later than mid-afternoon.
4. Unwind: Wind down to bedtime. Stop all work and stressful activities about an hour and a half before going to bed.
5. Go to bed early only when you are sleepy or tired. Do not go because of habit, and certainly not just because of convention.
6. Lights out: As soon as you slip between the sheets, put the lights out.
7. Do not read or watch TV in bed.
8. Do relaxing exercises for the first 10-20 minutes after retiring.

9. Relax in bed and reflect on what sort of day it has been. Stress the positive and eliminate the negative.
10. When failing to fall asleep within the first 10-20 minutes, get up, get dressed and get involved in some activity outside the bedroom. When sleepy again, go back to bed and switch off the light.
11. Repeat step 10 if necessary.
12. Wake up! Set the alarm to wake you at the same time every day.

It has been said that if you are at peace with God, you will be able to relax and sleep like a baby.

Choosing Good Food

There are as many stories about nutrition and good food as there are writers. However, a successful wife knows that unrefined and unprocessed foods are best for the family. These include fruits, vegetables, whole grains, nuts, meat (lean) and salads.

At times it may be difficult to avoid processed foods when one has absolutely no time to go to the market.

Economic considerations also make it impossible to access natural foods depending on where one lives. In Kenya we are lucky to have fresh foods in the market all year round. A successful wife can take advantage of the available foods and plan her menu according to the seasons of availability.

Smoke and Toxic Fumes

Smoke and toxic fumes from cars and industries have been linked to various cancers in the body. A professor in a linguistics class remarked that we were trying to ban cigarette smoking in public places when we allowed diesel engine fumes to pollute the city. She pointed out that breathing in diesel fumes was equivalent to smoking one thousand cigarettes.

These fumes could also cause lead poisoning if the vehicles use leaded fuel. Lead can also find its way to our foods from the ink used in labelling food-packaging papers when they come into contact with moisture.

In some eateries people may use old newspapers for covering their foods or even wrapping for customers take away. This is as dangerous as using the empty packets of flour for wrapping foods. The ink used on all these contains lead and it can be highly toxic.

Lead poisoning can also be got from eating vegetables that have been planted by the roadside. These absorb fumes from cars and store up this lead.

Stress Management

There are events in our lives that may be a cause of stress for us but a source of joy for others. We all experience stress, but how deeply it affects us depends on our personality.

A successful wife understands the dynamics of stress and tries to keep it under control as much as possible. Having to balance housework and work for income generation can not only be challenging but also stressful. Putting every effort to rise to the top rung in the corporate ladder can be greatly stressful. Trying to balance all this with running the home can lead to more stress for the wife who loves her family.

Managing stress begins in the mind. The first step is to look at the causes of stress and try to rationalize. "Supposing I'm dead, what could happen to this situation I'm trying to cling onto and behave as if I were indispensable? In the office delegate as much as possible, so as to avoid too much pending work.

At home and work, reduce clutter and get your paperwork under control.

A tidy office makes you motivated to stay there longer.

The stress could be arising from marriage itself. Misunderstandings between husband and wife can cause so much stress that living together becomes impossible. A husband who decides to take another wife becomes impossible to live with. He has broken the covenant that the two of them made before God and therefore he is no longer the man she married. He is not there for her anyway. A successful wife does not abandon her home and children because of the misconduct of her husband. She continues to live there with her children but makes sure she has no sexual relationship with the husband.

She cannot allow herself to be shared with someone else in that respect. She refuses to imitate the animals where sex is not exclusive and can be shared.

If he leaves, she resolves to practise chastity and mortification.

There are many manuals on stress but what works for one person may not work for another. It is important to establish an exercise routine that gives the body an occasional workout and a daily activity.

CHAPTER 9

GROOMING AND CREATING HARMONY

Even the sloppiest of women wants to look her best. Being neat and presentable is a must for all. Hence we may try to buy clothes that strike us as beautiful as well as making us feel comfortable in them. When a woman looks her best, it is usually not due to her clothes only. Good looks involve taking care of the whole person. One has to look after the hair, the face, the figure, the hands and feet and finally the clothes. One has also to feel good.

Care of the above is but care of the externals. For one to feel and look good, the interior as well as the exterior aspects are important. To create harmony, one has to pay caring attention to all levels of life.

Harmony means different rhythms working together, for instance, in dressing do colours match? Is the dress appropriate for the occasion? What about the accessories? And the scent? What about interiorly?

When failing to create this harmony, the results can be disturbing when not embarrassing.

Consider:

> A beautiful girl worked for a leading law firm as a secretary. She dressed elegantly and was the last word in combining dressing items. She was also very good at her work. However, whenever she went close to anyone, they would back up and look the other way. Her breath was so foul that everyone kept distances. She suffered for it, until a friend offered to tell her and help her out of the situation.

Many times we may offend others not only with our breath or perfume but also by tactless and insensitive behaviour. What is sensitive and insensitive may differ from one culture to another. What is uncouth in some cultures may be polite in others.

Dress

Apart from texture, colour is important when choosing a dress. Some people like certain colours. Others choose clothes depending on affordability. Colour is in most cases the first thing we notice in a person.

Experts agree that colour affects us in more ways than one. The colour we wear might provoke anger or excitement in other people. To look good in clothes the colours need to be analysed. They may be warm, cool or just neutral.

Irrespective of colour or race, people also fall into these two categories: cool toned or warm toned. Cool toned people look their best in cool colour shades like navy, azure, deep pink, white, aqua green, icy pink, maroon, silver etc.

Warm toned people on the other hand look best in warm based colours like yellow, brown, peach, copper, rust, lime and combinations thereof.

Hair

The purpose of a hair style is to enhance the looks of the face by concealing irregularities and creating certain moods. Every face is unique and the hair style adds to this uniqueness. A successful wife looks for the style that enhances the looks of her face best and complements her personality. She never neglects her hair. She sets aside a day of the week to attend to her hair, hands and feet.

Hands and Feet

Hands and feet also need to be cared for to remain soft and supple. Although commercial and home-made creams and lotions keep hands and feet soft one needs to invest an hour every week for a pedicure and manicure. This can be done at the salon or at home depending on one's finances.

Spirituality

Max Ehrmanns, composer of the poem 'Desiderata' tells us to be at peace with our God whatever we conceive him to be. All Mainstream religions agree that apart from love, peace is important for one's health.

Soldiers may think of peace in terms of war, but a successful wife thinks differently. She knows that her peace is not brought about by the gun but by a special spiritual experience. Her spiritual experience is something unique and need not be achieved through organized religion. (Although many organized religions have a lot to offer in this area).

Her spirituality allows her to believe in a higher being that controls and gives her life. Many people have expressed dissatisfaction with organized religion up to becoming atheists, especially after unpleasant and unfortunate experiences with organized religion. A young fellow once asked me why Christ, Buddha, or Krishna should teach him how to relate with his God. I had only one answer: they all spent their entire lives communicating with this higher being and knew him better than those of us who have not had such an opportunity. They are therefore in a better position to tell us who he really is.

Take Maeya. She is a cheerful and loving woman when you meet her. She will not stop smiling and occasionally laughing until you leave – reason? What else is there to live for except to be happy?, she quips. She began practising religion when she was a small girl and has kept her faith ever since. She does not own much. For her, three dresses are an excess. Her attitude is, why have three when two will do?

She is 92, but she looks 60. Her contemporaries are either dead or cramped by ill health.

Her secret: laughter, love and peace from religion. To her everything has a purpose, even death. She says everyone has a good side, even a thief, and it is the good side she looks at.

Maeya was married. She has four children. Unfortunately her husband died in the Second World War and she has not remarried. She lives alone in her house but her grandchildren live nearby. She is usually busy with basket weaving when she is not working in her garden. She keeps some free-ranging chicken and two cats. She talks to them as if they were people.

Maeya has not forgotten the dance rhythms of her youth. She easily breaks into song and dance and remembers the past as if it were yesterday.

A summary of the qualities of a successful wife

- She upholds family values
- She has a spiritual life
- She is hard-working
- She has a positive mental attitude
- She does not suffer of self-pity
- She loves unconditionally

- She respects the rights and property of others
- She is not violent
- She is nurturing
- She does not fight for recognition or for position as head of the family
- She does not try to control her husband
- She is frugal with money
- She is orderly and has a plan of life
- She endures hardships and trials
- She is optimistic
- She wants the best for her family
- She plans for tomorrow
- She is always trying to learn more
- She is up to date with new developments in her field of expertise.

END NOTES

1. Readers Digest Association, Inc., *Guide to Medical Cures and Treatments,* 1996, p. 66.
2. Berkov Robert et al (eds.), *Merck Manual of Medical Information,* NY: Pocket Books, 1997, p. 476.
3. Kalellis Peter M., *Restoring Relationships in the Family and Outside the Family*, NY: Crossroad, 2001, p. 180.
4. Isaacs David, *Character Building: A Guide for Parents and Teachers*, Nairobi: Four Courts Press, 1976, p. 63.
5. James Baldwin, *Nobody Knows My Name,* NY: Dell Publishing, 1986. p. 469.
6. Wood Samuel E. & Wood Ellen Green, *The World of Psychology,* USA: Allyn & Bacon, 1994, p. 253.
7. Op cit..
8. Mehrotra Vivek, *Why My Horse Doesn't Drink,* New Dehli: Viva Books, 2006, p. 97.
9. Wood Samuel E. & Wood Ellen Green, *The World of Psychology,* USA: Allyn & Bacon, 1994.
10. Ibid.

11 Kalellis Peter M., *Restoring Relationships in the Family and Outside the Family*, NY: Crossroad, p. 72.

12 *Social Message of Christ,* p. 114.

13 Wood Samuel E. & Wood Ellen Green, *The World of Psychology,* USA: Allyn & Bacon, 1994, p. 353.

14 Ibid., p. 411.

15 See Diane Baumrind, *Methods of Parenting,* 1991.

16 Aronson Elliot et al. *Social Psychology: The Heart and the Mind.* USA: Harper Collins, 1994, p. 385.

17 Rinaldo Ronzani, *Saints of the Liturgical Year,* Nairobi: Paulines Publications Africa, 2006, p. 74.

FURTHER READINGS

LET US TALK ABOUT LOVE

Parent and Teenagers Talk about Love

By Raymond Boisvert

ISBN 9966-21-346-3; 56 pages; publication 1998

The author has endeavoured to let parents and teens know that it is actually possible and extremely important for them to have a dialogue that will benefit the teenagers and harmonize relationships in the family.

WHEN IT IS RIGHT TO SAY NO

Solving Problems in Relationships

By Guarrella-G. M. Sofia

ISBN 9966-08-107-0; 120 pages; publication 2005

The consequences of a no, spoken or unspoken, accepted or rejected, if well learned, determine the quality of relationships one will have in adulthood with others as well as with oneself.

BRINGING UP CHILDREN

In a Permissive Society

By Pamela Wanda

ISBN 9966-08-246-8; 120 pages; publication 2007

The family is the first school and the parents are the first educators of their children. The pattern of behaviour that each child learns and follows is in the hands of the parents. Bringing up children requires time, lots of it.

LOVE AND CONFLICT IN MARRIAGE

Handling Misunderstandings

By Mary Kibera

ISBN 9966-08-242-5; 112 pages; publication 2007

What are some of the ingredients neccesary to make a happy marriage and to make it work when love by itself is not enough? More than ever before in our society, a great number of young couples do not only contemplate quitting their marriage but already have. Why? ... Marriage is not a human invention, but a divine institution. It is a union of one man with one woman in total mutual self-giving and commitment for life.